Opportunities with Exotic Beams

Argonne National Laboratory, USA

4 – 7 April 2006

Proceedings from the Institute for Nuclear Theory – Vol.15

Proceedings of the Third ANL/MSU/JINA/INT RIA Workshop

Opportunities with Exotic Beams

editors

Thomas Duguet
Michigan State University, USA

Henning Esbensen
Argonne National Laboratory, USA

Kenneth M Nollett
Argonne National Laboratory, USA

Craig D Roberts
Argonne National Laboratory, USA

World Scientific

NEW JERSEY • LONDON • SINGAPORE • BEIJING • SHANGHAI • HONG KONG • TAIPEI • CHENNAI

Published by

World Scientific Publishing Co. Pte. Ltd.
5 Toh Tuck Link, Singapore 596224
USA office: 27 Warren Street, Suite 401-402, Hackensack, NJ 07601
UK office: 57 Shelton Street, Covent Garden, London WC2H 9HE

British Library Cataloguing-in-Publication Data
A catalogue record for this book is available from the British Library.

Cover Art: Density profiles 150 microseconds after helium detonation on the surface of an accreting neutron star. The detonation started at the lower left and propagated across the 2 km width of the simulated region, producing surface waves and launching a photosphere that could rise 10 km above the surface. The *rp*-process might be triggered in similar events. This snapshot is from a calculation performed at the ASC Center for Astrophysical Thermonuclear Flashes at the University of Chicago and presented in M. Zingale *et al.*, Astrophysical Journal Supplement Series, Vol. 133, pp. 195–220 (2001).

OPPORTUNITIES WITH EXOTIC BEAMS
Proceedings of the Third ANL/MSU/JINA/INT RIA Workshop

ISBN-13 978-981-270-567-9
ISBN-10 981-270-567-8

Printed in Singapore by World Scientific Printers (S) Pte Ltd

SERIES PREFACE

The National Institute for Nuclear Theory Series

The national Institute for Nuclear Theory (INT) was established by the US Department of Energy in March, 1990. The goals of the INT include:

1. Creating a productive research environment where visiting scientists can focus their energies and exchange ideas on key issues facing the field of nuclear physics, including those crucial to the success of existing and future experimental facilities;
2. Encouraging interdisciplinary research at the intersections of nuclear physics with related subfields, including particle physics, astrophysics, atomic physics, and condensed matter;
3. Furthering the development and advancement of physicists with recent Ph.D.s;
4. Contributing to scientific education through graduate student research, INT summer schools, undergraduate summer research programs, and graduate student participation in INT workshops and programs;
5. Strengthening international cooperation in physics research through exchanges and other interactions.

While the INT strives to achieve these goals in a variety of ways, its most important efforts are the three-month programs, workshops, and schools it sponsors. These typically attract 300-400 visitors to the INT each year. In order to make selected INT workshops and summer schools available to a wider audience, the INT and World Scientific established the series of books to which this volume belongs.

In January 2004 the INT and several partners, Argonne National Laboratory, Michigan State University, and the Joint Institute for Nuclear Astrophysics, began a new workshop series to explore scientific questions that might be answered by the proposed Rare Isotope Accelerator. This volume summarizes the proceedings of the third workshop in this series, which was hosted by Argonne National Laboratory in April 2006 and organized by Thomas Duguet, Henning Esbensen, Ken Nollett, and Craig Roberts, the editors of this volume. The organizers designed a scientific program to provide a broad overview of rare isotope experiments and their implications for theory, including nuclear structure, dynamic symmetries, nuclear astrophysics, and nuclei as laboratories for testing fundamental interactions. The intent was to provide a broad summary of RIA science. The workshop proceedings are being published so that this overview will be available to the broader nuclear physics community, as the planning for this new facility progresses. As series editors, we would like to thank Argonne for graciously hosting this meeting and the organizers for the considerable effort they invested in designing the scientific program and in editing this volume.

This volume is the 15th in the INT series. Earlier series volumes include the proceedings of the 1991 and 1993 Uehling summer schools on Nucleon Resonances and Nucleon Structure and on Phenomenology and Lattice QCD; the 1994 INT workshop on Solar Modeling; the tutorials of the spring 1997 INT program on Tunneling in Complex Systems; the 1998 and 1999 Caltech/INT workshops on Nuclear Physics with Effective Field Theory; the proceedings of the 1998 RHIC Winter Workshop on Quarkonium Production in Relativistic Nuclear Collisions; the proceeding of Nucleon Resonance Physics (1997), Confinement III (1998), Exclusive and Semi-exclusive Reactions at High Momentum (1999), Chiral Dynamics 2000, and the Phenomenology of Large-N QCD (2002), all collaborative efforts with Jefferson Laboratory; the 2004 workshop on the Astrophysical Origin of the Heavy Elements, the first in the new Rare Isotope Accelerator series; and the 2004 workshop on Open Issues in Core Collapse Supernova Theory. We intend to continue publishing those proceedings of INT workshops and schools which we judge to be of broad interest to the physics community.

Wick C. Haxton and Ernest Henley
Seattle, Washington, August, 2006

PREFACE

The Third ANL/MSU/JINA/INT RIA Theory Meeting was held at Argonne National Laboratory from 4 – 7 April, 2006. It was one of a series of workshops that are hosted in a rotating fashion among the sponsoring institutions. The series was established to explore, explain and support the case for an advanced exotic-beam facility from a theoretical-physics perspective, and to encourage the theoretical work that will be needed to interpret experiments once such a facility is built.

An advanced exotic-beam facility has been a recognized priority of the US nuclear physics community since the 1996 Long-Range Plan formulated by the Nuclear Science Advisory Committee (NSAC). It is vital for the future of low-energy nuclear physics and will produce advances: in understanding the nature of nuclei and nuclear matter; in explaining the origins of the chemical elements and the workings of stars; and in testing the standard model of particle physics. During the period since the Long Range Plan the need for such a facility has been reiterated many times by NSAC, by the broader nuclear physics community, and by the Department of Energy (DOE). Moreover, although the Rare Isotope Accelerator (RIA) concept was materially modified shortly before this third meeting took place, both the community and DOE remain firmly committed to such a facility. Indeed, the charge letter requesting NSAC to prepare a new long range plan states that "...the projected funding for DOE is compatible with [...] commencing construction of the proposed Rare Isotope Accelerator early in the next decade."

The first two meetings in this series focused on specific areas of nuclear theory. The topic of the first, held at the Institute for Nuclear Theory (INT), was the astrophysical r-process. The second, held at Michigan State University (MSU), concentrated on the present state of nuclear reaction theory. Owing to the timing of this third meeting, we wanted the presentations, the discussions and, indeed, the proceedings volume to cover a broader range of topics in current nuclear theory. Sessions were organized on: dynamic symmetries in nuclei; fundamental symmetries; nuclear structure; and nuclear astrophysics. With broad consultation, including input from Francesco Iachello and Wick Haxton, we recruited a collection of speakers to give a thorough overview of topics at the leading edge of nuclear theory. In addition, we encouraged participation from the wider nuclear physics community, both experiment and theory. There were sixty-six registered participants, and we are grateful to all for their involvement and for making an effort to address the issue we identified; namely: "What questions does your research pose that only an exotic beam facility can answer?"

The picture which emerged from the presentations is that progress is being made in a broad range of distinct but overlapping areas within nuclear theory. This is captured in the contributions to this volume, which span a wide range of topics, from the lightest nuclei to some of the heaviest, from *ab initio* to phenomenological approaches, from "pure" nuclear physics to challenging astrophysical problems that require proper accounting for nuclear properties. An expansion of the range of measurable nuclear properties beyond present capabilities will present new constraints and new challenges for the theoretical approaches discussed in each of the chapters in this volume.

We set aside thirty minutes at the end of each session for discussion in an open forum; these discussions were often lively. Such discussion, with broad representation from across the nuclear physics community, is important to get theory launched on a trajectory that will both support and profit from experiments at the next generation of experimental facilities. It is our hope that the conversations begun at Argonne have continued over the subsequent months in university corridors and conference centers, and that the present volume of proceedings from the workshop will help that discussion evolve into fruitful action.

The Third ANL/MSU/JINA/INT RIA Theory Meeting was supported by a grant from Argonne's Laboratory Director, Robert Rosner, and by

Department of Energy, Office of Nuclear Physics, contract no. W-31-109-ENG-38. Finally, and gratefully, the organizers take this opportunity to thank Debbie Morrison for her unstinting effort and enterprise during all stages of the planning, organization and operation of the meeting.

Henning Esbensen
Kenneth Nollett
Craig Roberts
Thomas Duguet

Argonne (Illinois)
and East Lansing (Michigan)
August 2006

CONTENTS

COUPLED-CLUSTER THEORY FOR NUCLEI *

D. J. DEAN

Physics Division
Oak Ridge National Laboratory
Oak Ridge, TN 37831-6373 USA
E-mail: deandj@ornl.gov

In these Proceedings, I discuss recent developments and applications of coupled cluster theory to calculations of properties of medium-mass nuclei. I will report on results for both the closed-shell nucleus ^{16}O and its neighbors. I will also discuss future directions involving the implementation of a three-body force and resonant states into the coupled-cluster problem.

1. Current perspectives in the study of nuclei

Six years from now, new nuclear facilities will be churning out experimental data on short-lived nuclei. RIBF at RIKEN in Japan, SPIRAL-II, and GSI-FAIR will all be well on the way to either completion or full schedules of runs. The U.S. will also have an enhanced program established to continue research associated with the physics of nuclei. Today's experimental facilities are paving the way with exciting results concerning the nature of closed shells, the behavior of light nuclei near the drip lines, and the characteristics of very neutron-rich and unstable nuclei. Such results also impact our understanding of element production in the Universe, and have various application-oriented uses.

Nuclear theory during the same time frame faces the continuing challenges of developments that enable answers to the question(s): "Given a lump of nuclear material, what are its properties, where did it come from, and how does it interact?" We seek a unified framework for the description of nuclei and nuclear reaction processes that allows for accurate predictions and quantifiable error estimates for those predictions. Important in this pursuit is the understanding that we will not be able to measure every nuclear property that may be important even with the most sophisticated future experimental facilities. For example, neutron

*This research is supported by the Office of Nuclear Physics, Office of Science of the U. S. Department of Energy under Contract Number DE-AC05-00OR22725 with UT-Battelle, LLC (Oak Ridge National Laboratory).

cross sections on highly radioactive species may be important in the design of next-generation reactors or in various astrophysical nucleosynthetic processes, but their actual measurement may be difficult to impossible.

The nuclear many-body problem represents unique challenges and yet entails significant similarities with other fields of science. We use techniques to solve the many-body problem that are used in other quantum systems (for example, Green's function Monte Carlo, Hamiltonian diagonalization, many-body perturbation theory, coupled-cluster theory, Bloch-Horowitz, nuclear DFT and extensions to it). All of these methods focus on solutions of a problem filled with difficulty and yet at the same time, in some instances, possessing rather astonishing simplicity (magic numbers, for example). The nuclear quantum many-body problem has the unique feature of being driven by a Hamiltonian that is characteristically short-ranged, with many active operators (spin, angular momentum, isospin, tensor,...). In contrast to atoms and molecules, the nuclear interaction self-binds protons and neutrons in the nucleus. Furthermore, the nuclear interaction is not completely understood, although one may argue that it is, in fact, pretty much understood. Indeed recent progress enabled the community to describe certain nuclear properties through mass 12 in an ab initio many-body framework. The future holds the promise of using these ab initio results to guide one to the appropriate choice of density functional to be used in heavy nuclei calculations.

Theoretical challenges facing our field were outlined in the RIA Theory Blue Book[1] written in 2005 by a large part of the community. That document outlined broadly our efforts to describe nuclei. They include:

- development of ab initio approaches to medium-mass nuclei,
- development of self-consistent nuclear density-functional theory methods for static and dynamic problems,
- development of reaction theory that incorporates relevant degrees of freedom for weakly bound nuclei,
- exploration of isospin degrees of freedom of the density-dependence of the effective interaction in nuclei,
- development and synthesis of nuclear theory, and its consequent predictions, into various astrophysical models to determine the nucleosynthesis in stars,

These bullets represent the community's long-term goals for the pursuit of studies of nuclei. I might add a further applications-oriented bullet: development of robust theory and error analysis for nuclear reactions relevant to various security and energy applications.

In the remainder of these Proceedings, I will touch on work in coupled-cluster theory which is one of the methods tied to the first bullet in the list above. While we are currently developing coupled-cluster theory in closed-shell nuclei, our long-term goal is to develop the theoretical and computational tools of the coupled-cluster method that will be necessary to investigate very neutron-rich nuclei. We seek, with the application of coupled-cluster theory to nuclei, to describe some properties of larger nuclear systems in an ab initio framework. These developments represent one of the paths forward as we seek ultimately to answer the primary question "Given a lump of nuclear material what are its properties....".

2. Current progress in coupled-cluster theory for nuclei

Modern nucleon-nucleon interactions precisely reproduce phase shifts derived from experimental scattering data. They are based either on meson-exchange models[2,3] or generated through applications of effective field theory[4,5] to chiral Lagrangians that maintain the symmetries of QCD.[6] While these interactions reproduce nucleon-nucleon phase shifts up to a typical momentum scale of 350 MeV, they differ in their treatments of short-range interactions, and therefore results obtained from them for binding energies and nuclear spectra can be different. It has long been surmised[7] that a three-body interaction must be invoked to obtain appropriate nuclear binding, and effective field theory work indicates that three-body interactions arise naturally from diagrammatic power counting.[8] While based on slightly more phenomenological terms, the importance of nuclear three-body interactions for both ground-state energies and spectra was demonstrated by Green's Function Monte Carlo (GFMC) calculations.[9]

The GFMC approach[10] will be discussed in other proceedings of the Workshop. The advantage of the GFMC technique is its use of bare nuclear interactions. Its disadvantage, at least at present, is its incapability to treat interactions that are non-local in r-space. Other approaches rely on building a set of many-body basis states from the underlying single-particle basis. Using a basis necessarily leads to the need to renormalize the bare nuclear interactions for the given set of basis states. Furthermore, the above interactions all have a fairly repulsive core and therefore would require for direct implementation enormous basis sets to obtain converged results which capture the high-momentum components of the interaction (i.e., they would have to reach the 1 GeV scale). To avoid this, one resorts to methods that renormalize the interaction so that it may be computed in a small set of basis states. In standard notation, we divide the two-particle space into the P-space and Q-space such that $P + Q = 1$ and P and Q are projection operators. A goal is to make the P-space small enough to both capture the physics and allow for computation within that space.

Various approaches exist to determine the effective interaction for the P-space. We have used the Brueckner G-matrix approach for the renormalization that yields an effective interaction,[11,12] while the similarity-transform approach due to Lee and Suzuki[13,14] has been used in No-Core Shell Model (NCSM) calculations.[15] The advantage of the similarity-transform approach is that it yields a starting-energy independent solution for the two-body effective interaction, whereas the G-matrix contains the starting-energy (ω) as a parameter, since it is the solution to an equation of the form $G(\omega) = V_{NN} + V_{NN}\frac{\tilde{Q}}{\omega - t}G(\omega)$, where $\tilde{Q}$ allows only two-body state scattering, t describes the one-body part of the Hamiltonian, and V_{NN} is the bare two-body interaction. In many of our previous coupled-cluster calculations, we used the G-matrix with the Bethe-Brandow-Petschek theorem[16] to alleviate much of the starting-energy dependence. Quantitatively, ^{16}O calculations indicate that a 20-MeV change in the starting energy results in a 2-MeV change in the total binding energy using this prescription.

Once we have an effective interaction in a given model space, we then need to set up a many-body technique to solve for the correlated wave function, energy levels, and transition operators within a nucleus. One approach to the problem would be to diagonalize the renormalized nuclear Hamiltonian within the set of basis states as is done in the NCSM.[15] Please note, NCSM is a many-body theory, not a model and has been rather successful in describing light nuclear data.

An alternative approach to Hamiltonian diagonalization is coupled-cluster theory[17,18,19,20] which represents the wave function as an exponential rather than a linear combination of particle-hole excitation operators acting on a reference Slater determinant. In coupled-cluster theory the correlated many-body wave function is

$$\mid \Psi\rangle = \exp(T) \mid \Phi\rangle \ , \tag{1}$$

where $\mid \Phi\rangle$ is a Slater determinant representing the non-interacting system and $T = T_1 + T_2 + ...$, with T_n being an n-particle-n-hole correlation operator of the form

$$T_n = \sum_{ab...,;ij...} t_{ij...}^{ab...} a_a^\dagger a_b^\dagger \cdots a_j a_i \ . \tag{2}$$

The amplitudes may be determined by solving the non-linear coupled algebraic equations

$$\langle \Phi_{ij...}^{ab...} \mid \exp(-T) H \exp(T) \mid \Phi\rangle = \langle \Phi_{ij...}^{ab...} \mid [H\exp(T)]_C \mid \Phi\rangle = 0 \ , \tag{3}$$

where H is the Hamiltonian, C means only linked diagrams enter, and $\mid \Phi_{ij...}^{ab...}\rangle$ are excited Slater determinants built upon the uncorrelated ground-state determinant $\mid \Phi\rangle$. Labels $a, b, c, \cdots$ represent particle states while $i, j, k, \cdots$ represent hole states. If we work within the coupled-clusters in singles and doubles (CCSD) framework, then one simultaneously solves for the t_i^a 1p-1h and t_{ij}^{ab} 2p-2h amplitudes from Eqn. 3, while all other higher-order amplitudes are assumed to be zero. For two-body interactions, the similarity transformed Hamiltonian is given by

$$\bar{H} = (H\exp(T))_C = \left(H + HT_1 + \frac{1}{2}HT_1^2 + HT_2\right)_C . \tag{4}$$

Note that although we solve only for the 1p-1h and 2p-2h amplitudes, the wave function contains many higher-order excitations due to the exponential character of the correlation operator $\exp(T)$.

We use Hamiltonians of the form

$$H = T - T_{\rm CoM} + V + \beta_{\rm CoM} H_{\rm CoM} , \tag{5}$$

where V is the two-body interaction, $T_{\rm CoM}$ is the center-of-mass kinetic energy, T is the kinetic energy of the particles, and $\beta_{\rm CoM} H_{\rm CoM}$ is a correction term applied such that the expectation value $\langle H_{\rm CoM}\rangle = 0$ at some (on the order of 0.01 in larger model spaces) value of $\beta_{\rm CoM}$.

In several papers[21,22,23,24] we have developed the coupled-cluster techniques for nuclear systems. Our initial results have all been obtained using a G-matrix formalism for model spaces including up to 8 major oscillator shells. We found that the oscillator energy ($\hbar\Omega$) dependence of results decreases significantly around the $\hbar\Omega$ minimum as one increases the model space. We also found convergence of results with model-space size. We showed that for the ^{16}O system, non-iterative triples corrections hardly affect the total binding energy, and using the equation of motion CCSD (EOMCCSD) method,[25,26] we calculated the excitation energy of the first excited 3^- state to be 11.5 MeV (and with excited-state non-iterative triples corrections, 12.0 MeV).

Furthermore, we recently completed a study of the mass $A = 15, 17$ nuclei around ^{16}O using particle-removed and particle-attached EOM methods.[24] We illustrate the results of these calculations for 8 major oscillator shells (480 single-particle basis states) in Table 1. These results illustrate the coupled-cluster capability we have developed during the last three years.

As shown in Table 1, the CD-Bonn and the N^{3}LO models result in the largest spin-orbit splittings (much larger than in the case of V_{18}). While we used the $\hbar\Omega$ which yields the minimum binding energy for ^{16}O at $N = 8$ oscillator shells, and the ^{16}O ground-state has stabilized, there remains some $\hbar\Omega$ dependence in these excited states. For example, for $N = 6$, the excited states of the $A = 15$ system change by approximately 0.4 MeV in the $\hbar\Omega = 11 - 14$ MeV window, while the $A = 17$ states change by approximately 0.7–1.1 MeV in the same $\hbar\Omega$ window. This would indicate a stronger dependence on $\hbar\Omega$ for the $A = 17$ nuclei than

Table 1. A comparison of the energies of the low-lying excited states of ^{15}O, ^{15}N, ^{17}O and ^{17}F, relative to the corresponding ground-state energies (the $(1/2)_1^-$ states of ^{15}O and ^{15}N and the $(5/2)_1^+$ states of ^{17}O and ^{17}F), obtained with the PR-EOMCCSD (^{15}O and ^{15}N) and PA-EOMCCSD (^{17}O and ^{17}F) methods, the N^{3}LO[5], CD-Bonn[3], and V_{18}[2] potentials, and eight major oscillator shells, with the experimental data in the last column[27]. All entries are in MeV. For the CD-Bonn and N^{3}LO interactions, we used $\hbar\Omega = 11$ MeV. For V_{18}, we used $\hbar\Omega = 10$ MeV. For eight major shells $\beta_{\mathrm{CoM}} = 0.0$.

	Interaction			
Excited state	N^{3}LO	CD-Bonn	V_{18}	Expt
^{15}O $(3/2)_1^-$	6.264	7.351	4.452	6.176
^{15}N $(3/2)_1^-$	6.318	7.443	4.499	6.323
^{17}O $(3/2)_1^+$	5.675	6.406	3.946	5.084
^{17}O $(1/2)_1^+$	-0.025	0.311	-0.390	0.870
^{17}F $(3/2)_1^+$	5.891	6.677	4.163	5.000
^{17}F $(1/2)_1^+$	0.428	0.805	0.062	0.495

for the states in the $A = 15$ and ^{16}O nuclei. We see a similar, but decreasing, dependence in the $N = 7$ calculations, where we only performed the check at $\hbar\Omega = 14$ MeV. At least for the hole states we obtain results which stabilize as a function of the number of shells. For the excited states of ^{17}O and ^{17}F, there is still a relatively strong dependence on the number of shells and $\hbar\Omega$. The $(3/2)_1^+$ states are known resonances, and we do therefore expect that our approximation at the particle-attached EOMCCSD calculations may miss some important correlations in this case.

Using information we have for the N^{3}LO interaction, including the $A = 15, 17$ results, and the result of EOM calculations that yield the first-excited 3^- state at 11.5 MeV of excitation, we are able to see the nature of the failure of the two-body interaction. Note that experimentally, the first excited 3^- state in ^{16}O lies at 6.1 MeV. In the lowest order approximation, this state should be a 1p-1h excitation from the $0p_{1/2}$ hole state to the $0d_{5/2}$ particle state relative to the ^{16}O ground state. The energy required to produce such an excitation is

$$\Delta\epsilon_\pi = \epsilon_\pi(0d_{5/2}) - \epsilon_\pi(0p_{1/2}) = +[\mathrm{BE}(^{16}\mathrm{O}) - \mathrm{BE}(^{15}\mathrm{N})] = 11.526\ \mathrm{MeV}, \quad (6)$$

for the proton case. Similarly for the neutron case, we obtain $\Delta\epsilon_\nu = 11.521$ MeV. Thus, experimentally, the interactions among nucleons lower the energy of the first excited 3^- state by $11.5 - 6.1 = 5.4$ MeV. For the N^{3}LO interaction, a similar argument yields $\Delta\epsilon_\pi = 15.846$ MeV and $\Delta\epsilon_\nu = 15.789$ MeV. Using our excitation energy of 11.5 MeV, we see that we are off by $15.8 - 11.5 = 4.3$

MeV. While this is an interaction- (and method) dependent result, it represents a fairly stable situation in our calculations. We find that the discrepancy of 4.3 MeV between theory and experiment for the energy gap between the $0p$ and $1s0d$ shells accounts for a large fraction of the missing 6 MeV needed to reproduce the first 3^- state of ^{16}O. In the future, we will investigate both the additional correlations brought in by higher-order coupled-cluster methods, and also the effect on the spin-orbit splitting and major shell splitting of three-body forces.

All of the above calculations were performed within the context of our implementation of the G-matrix.[12] Another avenue of defining the renormalized nuclear potential is known in the literature as $V_{\text{low}k}$.[28] The approach defines a low-momentum interaction (with momentum cutoff $\Lambda \sim 2.0$ fm^{-1}) by following a renormalization group (RG) equation from a very large cutoff to the desired low-momentum cutoff. The RG approach integrates out modes in the interaction with momenta larger than the cutoff. The procedure preserves the half-on-shell T-matrix scattering amplitudes, and therefore the phase shifts. Since the RG equation and Lee-Suzuki approaches are actually equivalent[29] in momentum space, there will necessarily be effective three-body forces that arise from following the RG approach to a low-momentum cutoff Λ.

Since $V_{\text{low}k}$ reproduces phase-shifts by construction, one can investigate whether it is a model-independent interaction. Using CCSD, I calculated the binding energy of ^{16}O using $V_{\text{low}k}$ derived from two different interactions, CD-Bonn and N^{3}LO, using the same cutoff $\Lambda = 2.0$ fm^{-1}. The results of these calculations are shown as a function of $\hbar\Omega$ in Fig. 1. One sees from this figure that the $V_{\text{low}k}$ associated with two different forces yields different binding. The difference is approximately 10 MeV, which compares to an approximately 14-MeV difference we see between the ground-state energies using our G-matrix interactions at the same starting energies.[24] We can conclude that while $V_{\text{low}k}$ may yield phase equivalence at the two-body level, there is still off-shell behavior that does not make the potential model independent. One also sees significant overbinding with $V_{\text{low}k}$ which must be cured by the further inclusion of a three-body force, as was recently done in the few-body systems.[30]

3. Future paths in coupled-cluster theory

For nuclei with realistic interactions, one should also include three-body interactions. Furthermore, even when one applies the Lee-Suzuki theory or the $V_{\text{low}k}$ approach to obtain an effective two-body interaction, one also necessarily generates an effective three-body potential, which one often ignores in calculations. The effects of this effective three-body potential should become larger as the single-

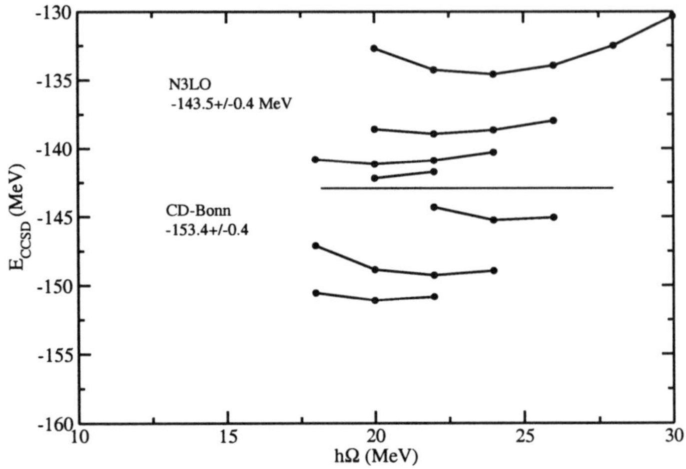

Figure 1. Comparison of $V_{\text{low}k}$ results for the CD-Bonn and N^{3}LO interaction at the same cutoff of $\Lambda = 2.0$ fm^{-1}. Lines indicate $N = 5, 6, 7, 8$ shell calculations for N^{3}LO and $N = 5, 6, 7$ shell calculations for CD-Bonn. An extrapolation to the infinite basis was made at using $N = 5, 6, 7$ data in both cases.

particle basis set becomes smaller. Furthermore, we know that real three-body forces should exist in nuclei. We are currently developing coupled-cluster theory at the CCSD level to incorporate three-body interactions.[31] Additional complications come from storage needs and the number of terms that enter Eq. 3. With a three-body interaction included in the Hamiltonian, we obtain for the coupled-cluster similarity-transformed Hamiltonian

$$\bar{H} = \left[H + HT_1 + \frac{1}{2}HT_1^2 + HT_2 + \frac{1}{6}HT_1^3 + HT_1T_2 + \frac{1}{24}HT_1^4 \right. \\ \left. + \frac{1}{2}HT_1^2T_2 + \frac{1}{2}HT_2^2 + \frac{1}{120}HT_1^5 + \frac{1}{6}HT_1^3T_2 + \frac{1}{2}HT_1T_2^2 \right]_C , \quad (7)$$

where clearly many more terms enter into the amplitude equations.

A second avenue that we are pursuing in coupled-cluster applications involves transforming the nucleon-nucleon interaction problem into a complex basis[32] through a complex rotation. Once the Hamiltonian is transformed to a complex basis, the coupled-cluster technology (also transformed to complex basis states and amplitudes) can then be used to investigate nuclear ground and resonant states.[33]

We are also beginning to work with computer scientists to construct the tensor-multiplies that are carried out within the coupled-cluster algorithm to scale to at least 20,000 processors. Our physics goal is to describe nuclei with 100 particles and 1,000 single-particle basis states. This will require an algorithm change to deal with memory and operation counts, but it is possible during a time when petascale computers will become available.

We should be able to report on each of the above activities in the next 3 years as we move towards ab initio calculations of medium-mass nuclei in the context of coupled-cluster theory. We believe this will be one tool to help us continue our quest to study the physics of nuclei.

Acknowledgments

Many people are involved in the coupled-cluster effort and results in these Proceedings could not have been obtained without their collaborative efforts. These include Morten Hjorth-Jensen (Oslo), Piotr Piecuch, Jeff Gour, and Marta Wloch (Michigan State University), Thomas Papenbrock (University of Tennessee and ORNL), and Gaute Hagen (ORNL).

References

1. See $http://www.orau.org/ria/RIATG/Blue_{B}ook_{F}INAL.pdf$
2. R. B. Wiringa, V. G. J. Stoks, and R. Schiavilla, Phys. Rev. C **51**, 38 (1995).
3. R. Machleidt, Phys. Rev. C **63**, 024001 (2001).
4. D. R. Entem and R. Machleidt, Phys. Lett. B **524**, 93 (2002).
5. D. R. Entem and R. Machleidt, Phys. Rev. C **68**, 41001 (2003).
6. C. Ordóñez, L. Ray, and U. van Kolck, Phys. Rev. C **53**, 2086 (1996).
7. J.G. Zabolitzky, K.E. Schmidt, and M.H. Kalos, Phys. Rev. C **25**, 1111 (1982).
8. E. Epelbaum, A. Nogga, W. Glöckle, H. Kamada, U.-G. Meissner, and H. Witała, Phys. Rev. C **66**, 064001 (2002).
9. R.B. Wiringa and S.C. Pieper, Phys. Ref. Lett. **89**, 182501 (2002).
10. S.C. Pieper, K. Varga, and R.B. Wiringa, Phys. Rev. C **66**, 044310 (2002).
11. M. Hjorth-Jensen, T.T.S. Kuo, and E. Osnes, Phys. Rep. **261**, 125 (1995).
12. D.J. Dean, T. Engeland, M. Hjorth-Jensen, M.P. Kartamyshev, and E. Osnes, Prog. Part. Nucl. Phys. **53**, 419 (2004).
13. K. Suzuki and S.Y. Lee, Prog. Theor. Phys. **64**, 2091 (1980).
14. K. Suzuki, Prog. Theor. Phys. **68**, 246 (1982); K. Suzuki and R. Okamoto, *ibid.* **70**, 439 (1983).
15. P. Navratil, G.P. Kamuntavicius, and B.R. Barrett, Phys. Rev. C **61**, 044001.
16. H.A. Bethe, B.H. Brandow, and A.G. Petschek, Phys. Rev. **129**, 225 (1963).
17. F. Coester, Nucl. Phys. **7**, 421 (1958).
18. F. Coester and H. Kümmel, Nucl. Phys. **17**, 477 (1960).
19. J. Čížek, J. Chem. Phys. **45**, 4256 (1966); Adv. Chem. Phys. **14**, 35 (1969).
20. J. Čížek and J. Paldus, Int. J. Quantum Chem. **5**, 359 (1971).
21. D.J. Dean and M. Hjorth-Jensen, Phys. Rev. C **69**, 054320 (2004).

22. K. Kowalski, D.J. Dean, M. Hjorth-Jensen, T. Papenbrock, and P. Piecuch, Phys. Rev. Lett. **92**, 132501 (2004).
23. M. Włoch, D.J. Dean, J.R. Gour, M. Hjorth-Jensen, K. Kowalski, T. Papenbrock, and P. Piecuch, Phys. Rev. Lett. **94**, 212501 (2005).
24. J.R. Gour, P. Piecuch, M. Hjorth-Jensen, M. Włoch, and D.J. Dean, Phys. Rev. C. in press (2006).
25. J. Geertsen, M. Rittby, and R.J. Bartlett, Chem. Phys. Lett. **164**, 57 (1989).
26. J.F. Stanton and R.J. Bartlett, J. Chem. Phys. **98**, 7029 (1993).
27. R. B. Firestone, V. S. Shirley, C. M. Baglin, S. Y. Frank Chu, and J. Zipkin, Table of Isotopes, 8th ed. (Wiley Interscience, New York, 1996).
28. S.K. Bogner, T.T.S. Kuo, and A. Schwenk, Phys. Rept. **386**, 1 (2003).
29. A. Schwenk, J. Phys. G: Nucl. Part. Phys. **31**, S1273 (2005).
30. A. Nogga, S.K. Bogner, and A. Schwenk, Phys. Rev. C **70**, 061002 (2004).
31. T. Papenbrock, G. Hagen, D.J. Dean, P. Piecuch, and M. Wloch, in preparation (2006).
32. G. Hagen, M. Hjorth-Jensen, and N. Michel, Phys. Rev. C, in press (2006).
33. G. Hagen, D.J. Dean, and M. Hjorth-Jensen, to be published.

THREE-BODY DECAY OF NUCLEAR RESONANCES

A.S. JENSEN, D.V.FEDOROV AND H. FYNBO
Department of Physics and Astronomy, University of Aarhus
NyMunkegade, DK-8000 Aarhus C, Denmark
E-mail: asj@phys.au.dk

E. GARRIDO
Instituto de Estructura de la Materia
Serrano 123, E-28006 Madrid, Spain
E-mail: e.garrido@iem.cfmac.csic.es

We use hyperspherical adiabatic expansion combined with complex scaling in computations of resonance energies and wavefunctions. We generalize the two-body decay of α-emission to three-body decay into three particles in the final state. The resonance position is reproduced by a three-body potential mocking up many-body effects at small distances. The width is determined by the adiabatic potentials at intermediate distances, whereas the final state energy distribution reflects the wavefunction at large distance. The characteristic features arising for large scattering lengths are discussed in relation with the Efimov effect in the continuum. Coulomb effects are treated in the same framework and tested on isobaric analog 2^+ states in $A = 6$ nuclei. Relatively strong isospin mixing is suggested to be a dynamic effect appearing outside the ranges of the short-range interactions where the Coulomb potential is decisive. Future kinematically complete experiments with high intensity beams are needed to constrain the theory.

1. Theoretical framework

The decay into three particles is undeniably a three-body problem at large distance. If the decaying initial state is an N-body system created in a resonance then the short-distance structure as undeniably is an N-body problem. As in the classic treatment of α-emission we shall artificially adjust the resonance energy to the desired value and otherwise treat all distances as a three-body problem[1]. This procedure can in principle only be posteriori justified but preliminary calculations indicate that intermediate and large distances are decisive for both width and final state properties[2].

We solve the three-body problem in coordinate space[3,4]. We use mass scaled

Jacobi coordinates $(\boldsymbol{x}, \boldsymbol{y})$ defined by

$$x^2 = r_{ik}^2 \, \frac{m_i m_k}{m(m_i + m_k)} \; , \quad y^2 = r_{j,ik}^2 \, \frac{m_j(m_i + m_k)}{m(m_i + m_k + m_j)} \; , \tag{1}$$

$$\rho^2 = x^2 + y^2 = \frac{1}{m(m_i + m_k + m_j)} \sum_{i<j} m_i m_j r_{ij}^2 \; , \tag{2}$$

where $\boldsymbol{r_{ij}}$ is the vector connecting particle i and j. The hyperspherical coordinates are then directions of $(\boldsymbol{x}, \boldsymbol{y})$ and relative size collectively denoted $\Omega = \{\Omega_x, \Omega_y, \alpha\}$, where $\tan\alpha = x/y$. Complex scaling amounts to the substitution $\rho \to \rho \exp(i\theta)$ in the equations.

In the adiabatic hyperspherical expansion we choose interactions and solve the Faddeev equations for each ρ providing angular eigenvalues λ_n and eigenfunctions $\Phi_n(\rho, \Omega)$ The three-body bound state or resonance wave function Ψ is then expanded on this complete set $\{\Phi_n(\rho, \Omega)\}$, i.e.

$$\Psi(\boldsymbol{x}, \boldsymbol{y}) = \sum_n f_n(\rho) \Phi_n(\rho, \Omega) \; , \tag{3}$$

where the radial wavefunction $f_n(\rho)$ is obtained in connection with the complex resonance energy.

2. Short-range interactions

The angular eigenvalues provide the adiabatic potentials as shown in Fig. 1 for an example where only short-range interactions are involved. At small distance the lowest potentials exhibit an attractive region followed by a barrier and a $1/\rho^2$ decrease towards zero. Such potentials can give rise to resonances where the energy is strongly depending on the depth and radius of the attractive pocket, the width is determined by the barrier and the final state energy distribution is related to the large-distance behavior of the wavefunction. The total wavefunction is found by a coupled channels computation involving the adiabatic potentials. The coupling potentials vary rather strongly for distances below 30 fm but all rapidly approach zero after ρ of about 40 fm. The corresponding radial wavefunctions, see eq.(3), shown in Fig. 2 oscillate around zero while falling off exponentially as functions of ρ. Their relative size at large distance determine the final state energy distribution. These ratios stabilize after ρ of about 40 fm.

The α-particle energy distribution found at large distance, see Fig. 3, receives contribution from several adiabatic wavefunctions. Interference is important. The data are relatively old and the calculations are not folded with experimental resolution. Still the high-energy peak is established and

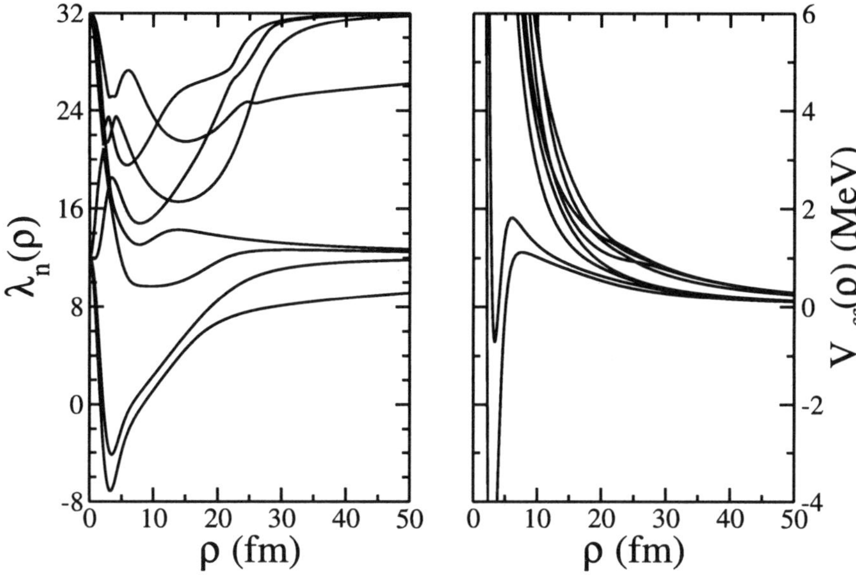

Figure 1. The real parts of the lowest 8 angular eigenvalues (left) and corresponding adiabatic potentials (right) as functions of ρ for the 2^+ states in ^{6}He (^{4}He + n + n). The scaling angle is $\theta = 0.10$.

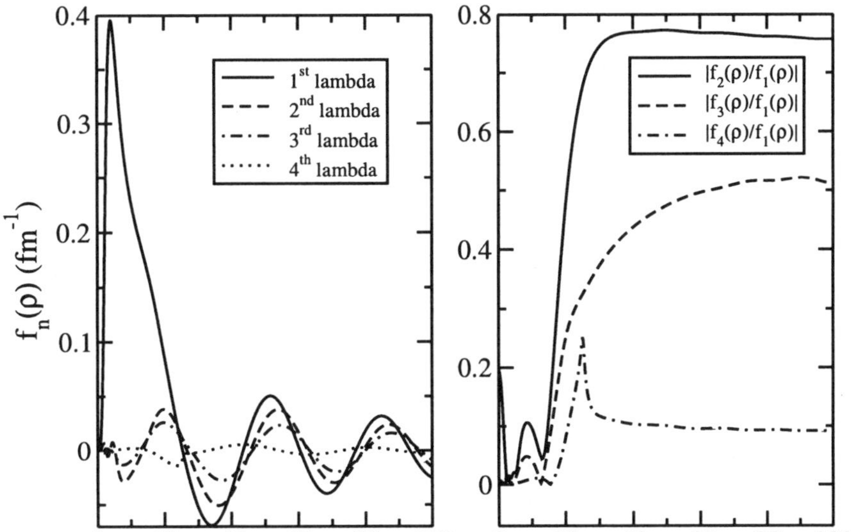

Figure 2. The lowest four radial wavefunctions (left) and their relative sizes (right) for $\theta = 0.10$ as functions of ρ for the ^{6}He(2^+) resonance.

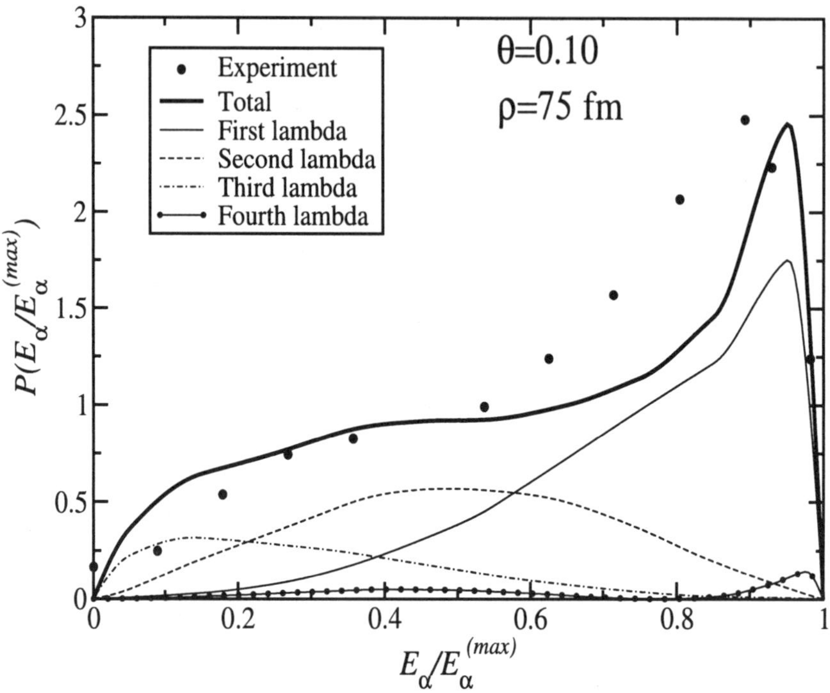

Figure 3. The energy distribution of the α-particle after decay of the 2^+-resonance in ^{6}He. The scaling angle is $\theta = 0.10$ and $\rho = 75$ fm where convergence is reached. The points are extracted from measurements[6]. Contributions from the lowest 4 adiabatic potentials are shown individually.

strikingly showing that the α-particle is emitted in opposite direction of two neutrons with very small relative momentum. The large neutron-neutron scattering length obviously provides sufficient attraction for this decay mechanism. It is similar to sequential decay but not produced by a stable intermediate configuration. It is then interesting to find the effect of more than one large s-wave scattering length[5]. A coherent sum of amplitudes should result precisely as for the bound Efimov states. This decay mechanism is investigated by use of a schematic model with only s-waves between all pairs of particles. The energy distributions in Fig. 4 are very stable against variation of the maximum distance ρ_{max}. The core energy has two peaks, i.e. one at high energy where the two neutron move together (intermediate neutron-energy peak) opposite to the core, and one at intermediate energy where one neutron is emitted with the largest part of the total energy (high-energy neutron peak), and the other neutron (low-energy neutron peak) follows the core.

These distributions are in Fig. 4 compared to the results from a realistic model where essentially all other known three-body properties of ^{11}Li are reproduced. The qualitative features essentially remain although less pronounced. The high-

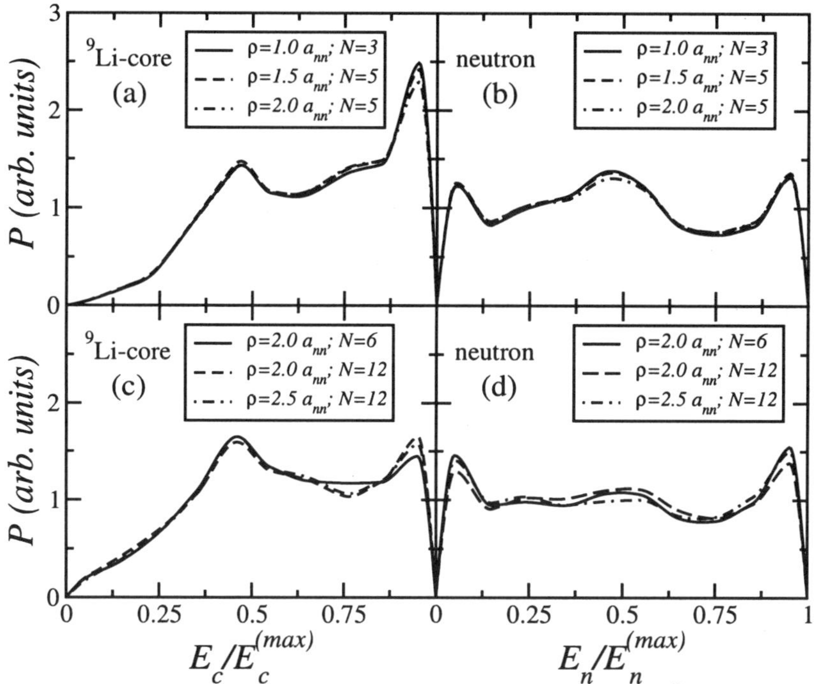

Figure 4. The energy distributions (upper parts) of the fragments - the core, ^{9}Li, and the neutrons - in the decay of a three-body resonance ^{11}Li(1^-) calculated in the three-body $^9\mathrm{Li}+n+n$ model with s-wave n-core interactions (scattering length $a_{nc} \approx 50$ fm), and an s-wave interaction in the $n-n$ subsystem (scattering length $a_{nn} \approx a_{nc} \approx 50$ fm). The different curves are calculated with different ρ_{max} and different numbers of adiabatic channels N to illustrate the convergence. The lower parts show the same for a realistic model[7] with more partial waves and n-core and n-n scattering lengths of about 20 fm.

energy core peak is reduced and consistently, arising from the same decay mechanism, the intermediate neutron energy peak becomes almost flat in a large energy region. The other peaks remain at the same energies and with the same relative heights, also arising from the same neutron-core intermediate configurations. The two coherent decay mechanisms, neutron or core emission, originate from the same effect of large scattering lengths. These features are then traces of the Efimov effect in the continuum.

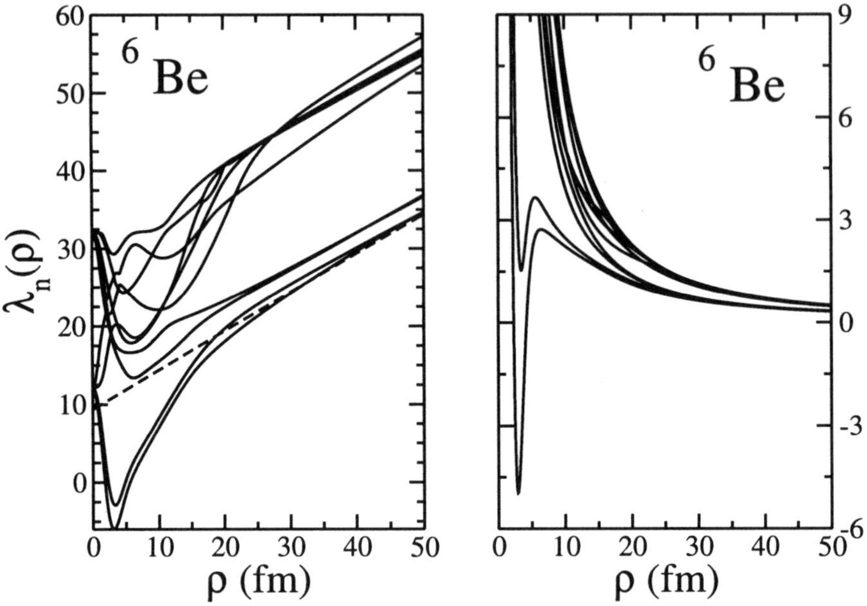

Figure 5. Real parts of the lowest 10 angular eigenvalues (left) and their corresponding adiabatic potentials (right) as functions of ρ for the 2^+-resonance in ^{6}Be. The scaling angle is $\theta = 0.15$. The dashed line is the estimated behaviour at large distances for the lowest angular eigenvalue.

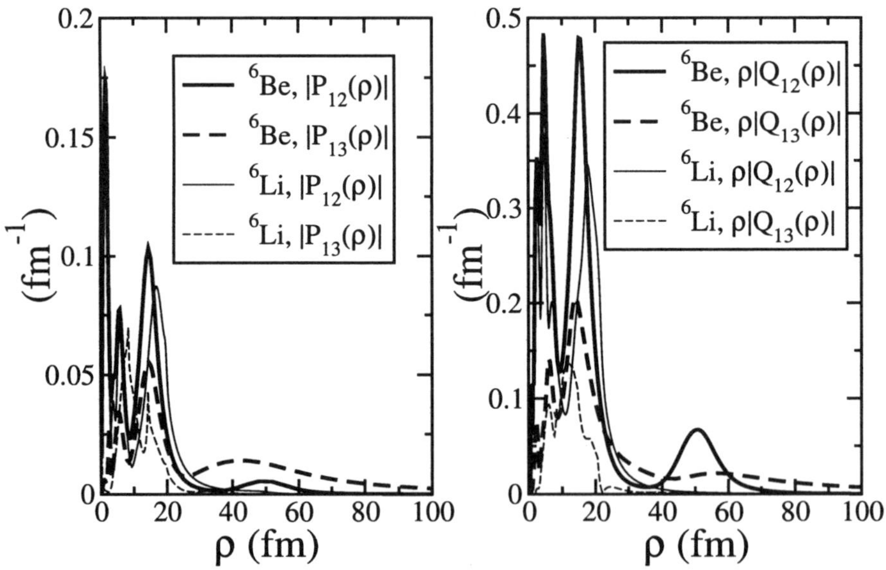

Figure 6. The absolute values of the coupling potentials between the three lowest adiabatic levels for the 2^+-resonance in ^{6}Be (thick curves $\theta = 0.15$ rads) as functions of ρ, and the corresponding isobaric analog states in ^{6}Li (thin curves $\theta = 0.10$ rads).

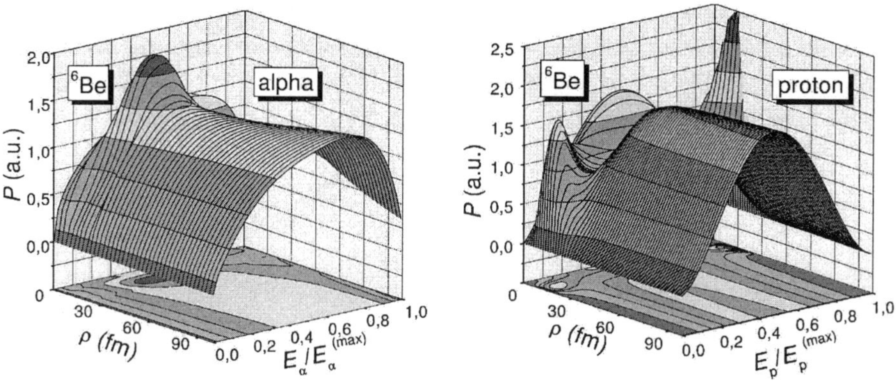

Figure 7. Kinetic energy distributions of protons (right) and α-particles (left) after decay of the 2^+-resonance in ^{6}Be. The three-dimensional plots show the dependence on ρ with inclusion of 10 adiabatic wave functions as function of $\cos^2\alpha$, i.e. the kinetic energies $E_{\alpha,p}$ are given in units of their maximum values $E_{\alpha,p}^{(max)}$, i.e. $(m_\alpha + m_p)/(m_\alpha + 2m_p)E_R$ and $2m_p/(m_\alpha + 2m_p)E_R$ for the proton and the α-particle, respectively, where E_R is the energy of the decaying resonance.

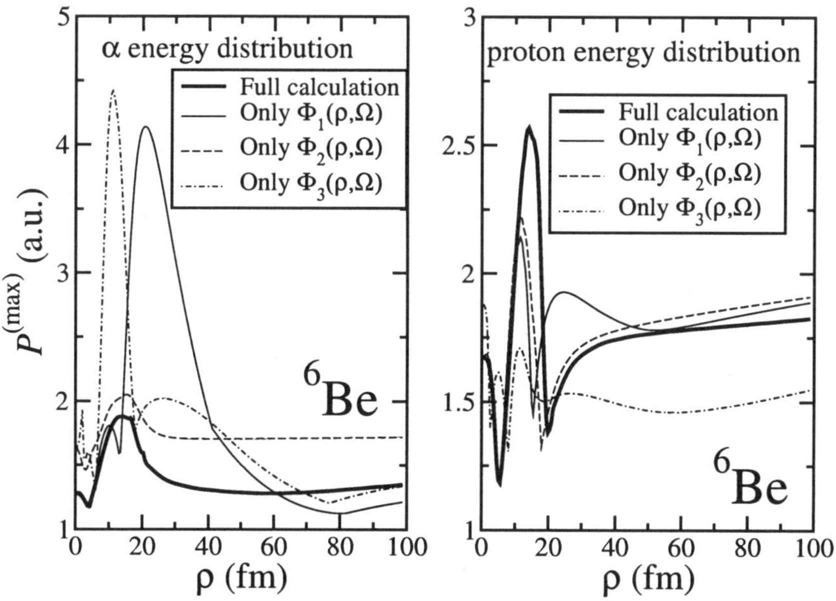

Figure 8. Projections of kinetic energy distributions for ^{6}Be-decay. Thick curves: Projection of the α (left panel) and proton (right panel) kinetic energy distributions (Fig. 7) on the $E_{\alpha,p}/E_{\alpha,p}^{(max)} = 1$ plane. They are then the profile originating from the maximum values of the energy distribution for each value of ρ. The thin curves are the same profile but when respectively only the first adiabatic term (solid), only the second adiabatic term (dashed), or only the third adiabatic term (dot-dashed) is included.

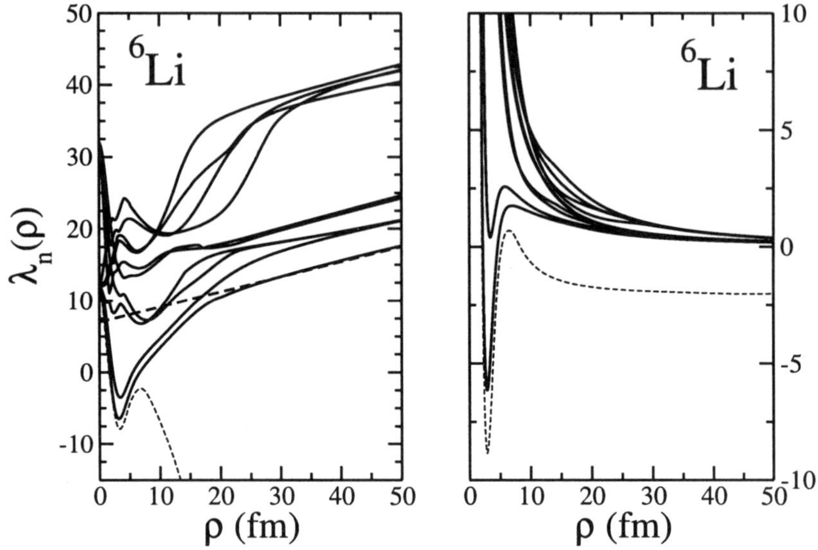

Figure 9. Real parts of the lowest 10 angular eigenvalues (left) and their corresponding adiabatic potentials (right) as functions of ρ for the 2^+-resonance in ^{6}Li. The scaling angle is $\theta = 0.10$. The dashed line is the estimated behaviour at large distances for the lowest angular eigenvalue.

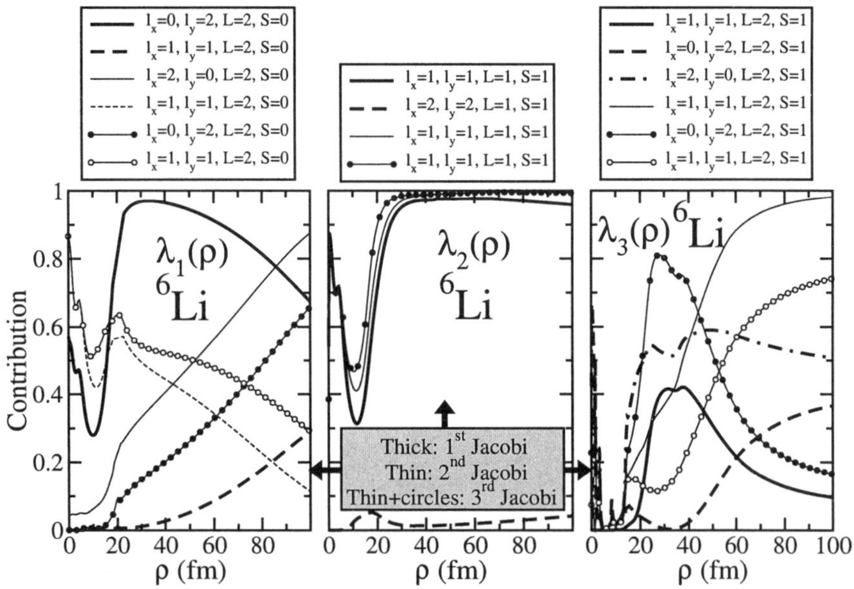

Figure 10. The fraction of the dominating components in the angular eigenfunction for the three lowest adiabatic potentials as function of ρ for ^{6}Li (2^+). The partial wave orbital angular momenta are (ℓ_x, ℓ_y, L) and S is the total spin. We omitted the almost decoupled lowest eigenfunction of deuteron-α character. In the second Jacobi set (thin lines) the x refers to the proton-α system and y to its center of mass motion relative to the neutron. In the third Jacobi set (thin+circle lines) the x refers to the neutron-α system and y to its center of mass motion relative to the proton.

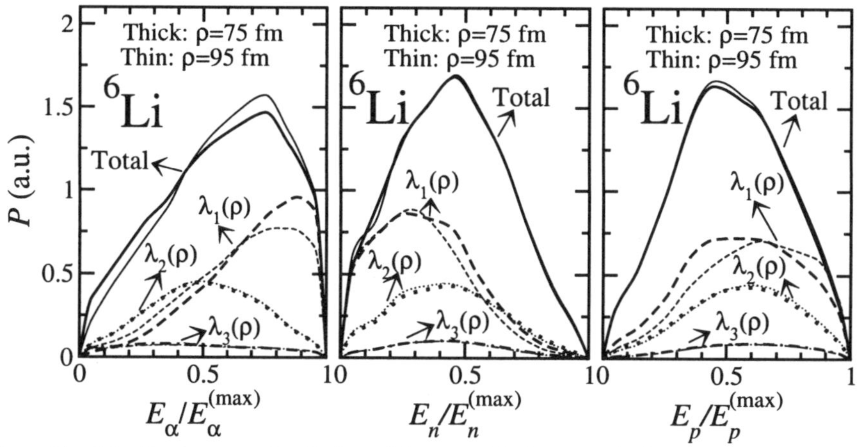

Figure 11. The kinetic energy distribution of the α-particle (upper part), the neutron (middle part) and the proton (lower part) after decay of the isobaric analog 2^+-resonance in ^{6}Li. The scaling angle is $\theta = 0.10$ and the two sets of curves are for $\rho = 75, 95$ fm. The points are extracted from the measurements[6]. Contributions from the lowest adiabatic potentials are shown individually.

3. Coulomb interaction and isopsin mixing for analog states

The Coulomb interaction couples adiabatic states at large distance. To test the method we investigate isobaric analog states, i.e. we first change the two neutron into protons and otherwise leave everything unchanged. The angular eigenvalues in Fig. 5 are then linear at large distance and the corresponding potentials decrease as $1/\rho$. The pockets are a little less attractive. The couplings in Fig. 6 vary at small distance but vanish at $40 - 60$ fm. This is promising for the numerical stability at large distance. In fact, the kinetic energy distribution in Fig. 7 is rather stable at large distance in contrast to the smaller distances. Clearly the resonance structure is changing as ρ increases. The protons peak at intermediate energies. The α-particles emerge in a broad peak tilted towards large energy. The effect of the Coulomb coupling is a broadening of the distribution and disappearance of the high-energy peak present for two neutrons in ^{6}He. The redistribution of the kinetic energy distribution with ρ is pronounced in Fig. 8 where the maximum probabilities of the total energy distributions are followed as ρ increases. The individual contributions from each adiabatic potential vary strongly whereas the total distribution is much more stable for both α-particles and protons.

The analog state in ^{6}Li has one neutron and one proton, i.e. the isospin zero components should also be included in the computations. This is seen in Fig. 9 where the downwards diverging eigenvalue corresponds to the bound deuteron channel. The corresponding lowest potential converges at large distance to a constant value equal to the deuteron binding energy. The slopes at large distance are now smaller than for ^{6}Be because of one charge less. Otherwise the adiabatic potential are as for the analog states. The partial wave decomposition of the angular wavefunctions are shown in Fig. 10. Again we see the strong variation with ρ for the different adiabatic states. The neutron-proton s-waves vary but dominate at all distances. The neutron-proton p-waves increase with ρ, and α-proton d-waves dominate at large distance. The isospin zero component of the first adiabatic wavefunction is very small at small distance but an increase sets in at around 50 fm. The corresponding quantum numbers are not consistent with the bound deuteron state but clearly the neutron-proton relative state in the continuum can be both isospin zero and one. These computations indicate that the isospin mixing is very small at small distance but at distance where the short-range interactions are negligibly small the Coulomb interaction strongly begins to mix the different isospin. The eventual outcome at large distances is at the moment not numerically sufficiently reliable to draw definite conclusions. However, this dynamical mixing seems to be very plausible.

The energy distributions, shown in Fig. 11, are consistent with measured data. We have, as for ^{6}He, contributions from several adiabatic components. The protons peak at intermediate (towards higher) and the neutrons (towards lower) energies. The α-particles appear in a broad peak leaning towards the high-energy side somewhat reminiscent of ^{6}He. The Coulomb interaction broadens the distribution but the neutron-proton s-wave attraction is visible.

References

1. E. Garrido, D.V. Fedorov, and A.S. Jensen, *Eur. Phys. J.* **A25**, 365 (2005).
2. E. Garrido, D.V. Fedorov, A.S. Jensen and H.O.U. Fynbo, *Nucl. Phys.* **A748**, 27 and 39 (2005), and **A766**, 74 (2006).
3. E. Nielsen, D.V. Fedorov, A.S. Jensen and E. Garrido, *Phys. Rep.* **347**, 373-459 (2001).
4. A.S. Jensen, K. Riisager, D.V. Fedorov and E. Garrido, *Rev. Mod. Phys.* **76**, 215 (2004).
5. E. Garrido, D.V. Fedorov and A.S. Jensen, *Phys. Rev. Lett.* (2006), in press.
6. B.V. Danilin, M.V. Zhukov, A.A. Korsheninnikov, L.V. Chulkov, V.D. Efros, *Sov. J. Nucl. Phys.* **46**, 225 (1987).
7. E. Garrido, D.V. Fedorov and A.S. Jensen, *Nucl. Phys.* **A708**, 277 (2002).

SOME CHALLENGES FOR NUCLEAR DENSITY FUNCTIONAL THEORY

T. Duguet

National Superconducting Cyclotron Laboratory and Department of Physics and Astronomy, Michigan State University, East Lansing, MI 48824, USA
E-mail: duguet@nscl.msu.edu

K. Bennaceur

Institut de Physique Nucléaire de Lyon, CNRS-IN2P3/Université Claude Bernard Lyon 1, 43, bd. du 11 novembre 1918, F-69622 Villeurbanne Cedex, France
CEA/DSM-DIF/ESNT, CEA/Saclay, 91191 Gif-sur-Yvette, France
E-mail: bennaceur@ipnl.in2p3.fr

T. Lesinski and J. Meyer

Institut de Physique Nucléaire de Lyon, CNRS-IN2P3/Université Claude Bernard Lyon 1, 43, bd. du 11 novembre 1918, F-69622 Villeurbanne Cedex, France
E-mail: lesinski@ipnl.in2p3.fr ; jmeyer@ipnl.in2p3.fr

We discuss some of the challenges that the DFT community faces in its quest for the truly universal energy density functional applicable over the entire nuclear chart.

1. Introduction

In the study of medium to heavy mass nuclei, nuclear Density Functional Theory (DFT), based on the self-consistent Hartree-Fock-Bogoliubov (HFB) method and its extensions, is the theoretical tool of choice. As new exotic beam facilities are being built or proposed to be built around the world, DFT is on the edge of becoming a predictive theory for all nuclei but the lightest. This is not only true for ground state properties, such as binding energies, radii or multipoles of the density, but also for low-energy spectroscopy and decay probabilities.

These advances are possible thanks to the development of better energy functionals and to the increase of computer resources. However, the needed accuracy and predictive power for unknown regions of the nuclear chart still leave a lot of room for improvements. Thus, DFT faces important challenges in its quest for the truly universal energy density functional which, in addition to allowing reliable calculations of extended nucleonic matter, nuclear masses, beta decay rates or fission lifetimes, is crucial to the understanding of the astrophysical r-process relevant to the nucleo-synthesis of heavy elements.

In the present contribution, some of those challenges are discussed[a]. As a matter of illustration, preliminary results of an ongoing study addressing the idea of using new *ab-initio* results to adjust the Skyrme energy functional are presented.

2. Challenges

The last decade has seen tremendous achievements in developing microscopic methods to describe medium to heavy nuclei. The most striking ones have come from nuclear DFT based on the self-consistent HFB method and its extensions (for a recent review, see Ref.[1]). Such methods rely on realistic effective interactions/functionals (either non-relativistic Skyrme or Gogny functionals, or relativistic Lagrangians) as the only (phenomenological) input. Specific correlations beyond the static mean field are now (almost) routinely incorporated through restoration of broken symmetries and configuration mixing calculations (projected-Generator-Coordinate-Method DFT).

Beyond the successes encountered in the development of the method and in the comparison with experimental data, several formal and practical challenges still have to be overcome in order to obtain a unified description of all systems throughout the chart of nuclei, in particular with the perspective of the construction of new radioactive beam facilities like RIA.

2.1. *Long term strategy*

Energy functional practitioners need a better-defined long term strategy in the spirit of the "Jacob's ladder" which is referred to in electronic DFT.[2] Indeed, constructing and adjusting functionals is a tricky game where it has been difficult to learn from other groups' work. Each group uses its own construction/fitting strategy, itself evolving in time in a way which is sometimes not systematic enough to allow for significant progress.

First, functionals at different levels of the many-body treatment must be consistent. For example, the HFB and the projected-GCM functionals must be connected in a clear way, the latter including the former as a particular case. Thus, projected-GCM DFT should be motivated from first principle and the HFB level

[a]This is *not* an exhaustive list of challenges. In particular, it does not cover all methods related to DFT, i.e. QRPA. Some elements discussed presently reflect the overall view of the DFT community and constitute the outcome of the workshop "Towards a universal density functional for nuclei", September 26-30, 2005, Institute of Nuclear Theory, Seattle, USA. Of course, some ideas discussed here reflect our personal view on the matter.

derived from it. This is still lacking. Second, any improved/complexified functionals should include all features validated by the previous/simpler ones. Third, a list of (truly) benchmark problems should be argued upon and used to make the energy functional consistent with ab-initio calculations of systems which are essentially solvable exactly; i.e. Brueckner-Hartree-Fock (BHF)[3] and variational chain summation (VCS) methods[4] in infinite matter, Green Function Monte Carlo (GFMC)[5] method and No Core Shell Model (NCSM)[6] for light nuclei, later followed by Coupled Cluster (CC) methods[7] for heavier systems. An example of that will be touched upon in section 3. Fourth, practitioners need to make progress regarding fitting algorithms. Systematic errors must be estimated and covariance analysis of parameters need to be performed to determine their relevant combinations.[8] Also, one has to assess the improvements brought about by further ingredients: new theory input and/or additional experimental data, i.e. superdeformed states, fission isomers/barriers of (exotic) nuclei. Finally, one can test new fitting algorithms (re-annealing,[9] genetics) to make sure that the absolute minimum in parameter space is reached during the fit. It is only following such a "strict" strategy that significant progress will be made in the near future.

2.2. *Improved phenomenology*

Improving single-particle energies coming out of the functional might be one of the most important subject of focus for the near future. Indeed, many unsatisfactory single-particle energy spacings provided at the HFB level spoil features of low-energy spectroscopy which cannot be corrected for by including further correlations through beyond mean-field techniques.[10] This has not only to do with the shell evolution towards the drip-lines but also with single-particle energies in stable systems. In any case, it is clear that data on particle/hole states in very neutron rich nuclei will help by providing shell evolution along long isotopic chains. For instance, information on particle/hole states around ^{78}Ni should be accessible with RIA.

However, there are issues as far as the fitting strategy is concerned. Indeed, one cannot fit single-particle energies directly to experiment without including their renormalization through the coupling to surface vibrations.[11] One would need to adjust functionals through beyond mean-field calculations of odd nuclei, which is too time consuming to be done in a systematic way within projected-GCM DFT. One could use other methods to estimate the effect.[11]

One has indications that the previous problem might be partly related to the omission of the tensor force in DFT. The latter could explain a lot of closing/opening of sub-shells as one fills major shells.[12] This is related to the attraction/repulsion between neutron (proton) $j = l + 1/2$ and proton (neutron)

$j = l - 1/2/l + 1/2$ orbits provided by the tensor interaction. The $N = 32$ sub-shell closure in neutron rich isotopes (^{54}Ti) is one possible example. Several groups are now working on implementing the tensor force in the Gogny and the Skyrme functionals.[13–15] It is important to note that the tensor force has been mostly disregarded in DFT so far because of a lack of clear-cut experimental data to adjust it. Again, observation of (relative) single-particle energies in very exotic nuclei could be of great help in that respect.

Masses, separation energies, densities, deformation, individual excitation spectra and collective excitation modes such as rotation or vibration, depend significantly on the superfluid nature of nuclei. The role of pairing correlations is particularly emphasized at the neutron drip-line where the scattering of virtual pairs into the continuum gives rise to a variety of new phenomena in ground and excited states of nuclei.[16]

Despite its major role, our knowledge of the pairing force and of the nature of pairing correlations in nuclei is rather poor. The structure of the functional as far as gradient versus (isovector) density dependencies are concerned has to be clarified.[16–18] Proper renormalisation techniques of local functionals have been recently proposed but their content must be further studied.[19,20] Also, particularly puzzling is the situation regarding beyond-mean-field effects associated with density, spin, isospin fluctuations.[21,22] Finally, the influence of particle-number projection and pairing vibrations has to be characterized through systematic calculations.

At this point, all modern phenomenological functionals provide similar pairing properties in known nuclei, i.e. odd-even mass staggering (OES) between ^{104}Sn and ^{132}Sn, while their predictions diverge in neutron rich systems,[23] where having masses up to $^{146/150}Sn$ or ^{81}Ni with an accuracy $\delta E = 50$ keV would already allow one to discriminate between most models. Pairing can be probed in other ways and data on high spin states, low-energy vibrations, two-nucleon transfer cross sections (feasible with beams of 10^4 part/s) in exotic nuclei will be complementary and all very valuable.

2.3. *Connection to underlying methods*

The energy functional has to be rooted into a more fundamental level, eventually connected to QCD. Doing so is not only important from a heuristic point of view but also from a practical one. Indeed, the main problem of current phenomenological functionals is the spreading in their predictions away from known nuclei. At the same time, phenomenology indicates that gradient and/or density dependencies of usual Skyrme functionals are too schematic/limited (see section 3.2). Thus, we need guidance beyond a fit on *existing* data. Hopes are put into deriving functionals from Effective Field Theory (EFT) which has been successfully applied to few-nucleon systems and which offers a comprehensive picture of the relative importance of three-body and two-body forces.[24] This could be done with the use of the renormalization group method in the spirit of V_{lowk}[25] from which a perturbative many-body problem seems to emerge in infinite matter.[26] The connection with finite nuclei will call for methods such as the Density Matrix Expansion.[27]

2.4. *Grounding beyond-mean-field DFT*

Ultimately, the functional has to incorporate, in a microscopic manner, long range correlations associated with large amplitude vibrational motion and with the restoration of broken symmetries. This is of particular importance to finite self-bound systems. The projected-GCM extension of DFT has proven to be one of the most promising ways of doing so. For the future, the first challenge will be to treat more collective degrees of freedom within the GCM, such as the pairing field or the axial mass octupole moment. So far, only the axial quadrupole moment has been considered in systematic GCM calculations.[1] Second, configuration mixing and symmetry restoration will have to be properly formulated, from first principle, within the context of DFT. While one can use guidance from the case of the average value of a Hamiltonian in a projected state, DFT presents key differences which have to be considered seriously. Because of the lack of theoretical foundation, one has relied on ad-hoc prescriptions so far to extend the HFB functional to the projected-GCM level, in particular as far as density dependencies in the functional are concerned.[28] Even worse, particle-number projected DFT is ill-defined as was recently realized[29] and fixed.[30]

3. The Skyrme energy functional : use of ab-initio inputs

In relation to section 2.1, we now briefly present a recent attempt to refine the $SLyx$ family of Skyrme functionals[31] by using additional ab-initio inputs from calculations of infinite nuclear matter. Our goal here is to highlight the general lessons one can learn from such an attempt.

3.1. *Constraining the isovector effective mass m^*_v*

Our original goal was to study the effect of the neutron-proton effective mass splitting $m^*_n - m^*_p$ with isospin asymmetry $I = (\rho_n - \rho_p)/\rho$ on the properties of the Skyrme functional. It is of importance since the effective mass controls the density of single-particle energies at the Fermi surface, and hence influences masses and shell correction in superheavy nuclei as well as how static and dynamic correlations develop.

While the isoscalar effective mass m_s^* (≈ 0.8) is well determined by the ISGQR,[32,33] constraints on the isovector one m_v^* ($\approx 0.7 - 0.9$) via the IVGDR are not decisive enough. On the other hand, recent BHF calculations of asymmetric infinite nuclear matter, with/without $3N$ force, predict $\Delta m_{n-p}^* \approx 0.22$ for $I = 1$[34] ($\Delta m_{n-p}^* \approx 0.13$ for Dirac BHF[35,36]).

Although the amplitude of the splitting cannot be considered as a ab-initio *benchmark*, its sign is more definite. Thus, we decided to construct Skyrme functionals with different values of Δm_{n-p}^* with the idea of improving their isovector properties by using predictions from ab-initio calculations. From that point of view, it is interesting to note that the $SLyx$ parameterizations were fitted to the Equation of State (EOS) of Pure Neutron Matter (PNM) with the idea of improving isospin properties of the functionals. One consequence was to generate functionals with $\Delta m_{n-p}^* < 0$, in opposition to ab-initio predictions. On the other hand, older functionals such as $SIII$[37] and SkM^*,[38] which were not fitted to PNM, had $\Delta m_{n-p}^* > 0$. The same exact situation happens for the Gogny force.[39] Thus, improving global isovector properties (EOS) seems to deteriorate state-dependent ones (m_v^*) with currently used functionals.

In any case, we tried to adjust three parameterizations (f_-, f_0, f_+) using the same fitting protocol as $SLy5$ but with different values of m_v^*. [b]. Their bulk properties in SNM are summarized in the table below. They are all identical and very close to SLy5', except for the effective mass splitting.

	ρ_{sat}	E/A_{sat}	K_∞	a_I	m_s^*	Δm_{n-p}^*
SkM*	0.160	-15.770	217	30	0.79	0.356
SkP	0.162	-15.948	201	30	1.00	0.399
SLy5'	0.161	-15.987	230	32	0.70	-0.182
f_-	0.162	-16.029	230	32	0.70	-0.284
f_0	0.162	-16.035	230	32	0.70	0.001
f_+	0.162	-16.036	230	32	0.70	0.170

First, a few lessons were learnt. Raising m_s^*/m to 0.8 is difficult because of spin-isospin instabilities. With $m_s^*/m = 0.7$, it is impossible to reproduce the high density part of PNM EOS and fix $\Delta m_{n-p}^* > 0$ at the same time[c]. It explains why the $SLyx$ forces predict the wrong sign of the effective mass splitting in neutron rich matter. Using two density-dependent terms in the functional $\propto \rho_0^{1/3}; \rho_0^{2/3}$[41]

[b]The only difference in the protocol is that static spin-isospin instabilities below $\rho = 2\rho_{sat}$ are excluded through the use of Landau parameters. SLy5' is as SLy5 except for the ab-initio EOS used in the fit; see Fig. 1.

[c]For all formula of relevant quantities, we refer the reader to.[40]

allowed the construction of (f_{-}, f_0, f_{+}).

In addition, $SLy5$-like functionals with $\Delta m^*_{n-p} > 0$ generate new finite-size isospin instabilities which become more dangerous as Δm^*_{n-p} becomes more positive. The instability tends to split neutron and proton densities by gaining energy through the term $\propto C_1^{\Delta\rho} \left(\vec{\nabla}\rho_1\right)^2$ in the functional, where ρ_1 is the scalar-isovector density. The table below gives typical values of $C_1^{\Delta\rho}$. We could check that SkP[42] already presents such an instability and that, empirically, $C_1^{\Delta\rho}$ must be greater than $\approx -30MeV$ to have an instability free functional. Those finite size instabilities need to be quantified and better controlled in the fit of the functional. For such a purpose, we have extended the control to finite wave-length instabilities by calculating the RPA response function in symmetric nuclear matter (SNM).[40,43]

	f_{-}	SLy5'	f_0	f_{+}	SkP
Δm^*_{n-p}	-0.284	-0.182	0.001	0.170	0.399
$C_1^{\Delta\rho}$	-5.4	-16.7	-21.4	-29.4	-35.0

The previous discussion shows already the type of problems/information arising from our attempt to improve on the fitting protocol of the Lyon functionals by using more ab-initio benchmark inputs. Now, the left panel of Fig. 1 shows PNM and SNM EOS as obtained for $(f_{-}, f_0, f_{+}$, SLy5') and as predicted by variational chain summation methods.[4] The asymmetry energy a_I as a function of density is displayed on the right panel. The message to extract from such a comparison is that the four forces have the ability to reproduce both microscopic EOS with the same accuracy. However, is that enough to characterize the spin-isospin content of the functionals and assess their quality? In order to answer such a question, we use further ab-initio predictions and look at the contributions to the SNM EOS per spin-isospin (S, T) channels.

3.2. *Separation of the EOS into (S, T) channels*

In this section, we discuss the contributions to the potential energy[d] of SNM from the four two–body (S, T) channels. We compare our results with those predicted by BHF calculations using Argonne V18 two-body interaction and a three-body force constructed from meson exchange theory.[45] By virtue of antisymmetrization, each spin–isospin channel is linked to a given parity of the spatial part of the effective force. However, and as motivated by the DME, the Skyrme functional

[d] The difference between the total energy and the kinetic energy of the corresponding free Fermi gas.

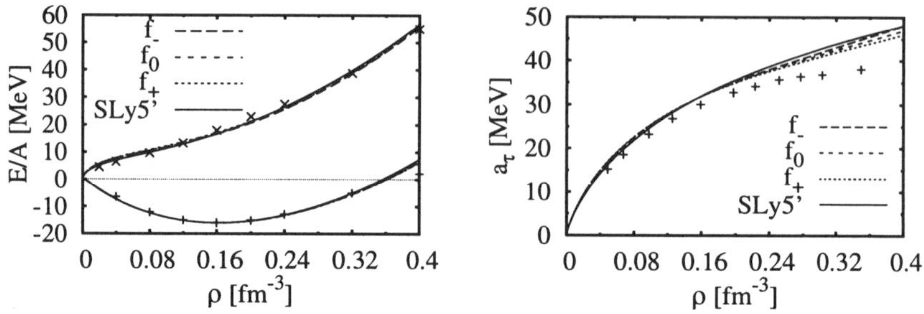

Figure 1. Left : SNM and PNM OES as given by functionals discussed here (see text), compared with VCS results by Akmal *et al.*[4] Right : isospin symmetry energy vs. density for the same set of functionals, compared to variational calculations by Lagaris and Pandharipande.[44]

is to be seen as the average of an effective vertex over the Fermi sea.[27] In particular, the angular dependence of the original nucleon-nucleon force have been averaged out. Thus, the fact that only S and P waves are explicitly considered in the Skyrme functional is illusory. In any case, one has access to the averaged spin-isospin content by looking at the different (S,T) contributions to the energy.

Using projectors on spin/isospin-singlet and spin/isospin-triplet states, it is possible to split the potential energy E_p into (S,T) contributions E_p^{ST}. In SNM, one finds :

$$\frac{E_p^{00}}{A} = \frac{3}{160}t_2(1-x_2)\left(\frac{3\pi^2}{2}\right)^{2/3}\rho^{5/3}, \tag{1}$$

$$\frac{E_p^{10}}{A} = \frac{3}{16}\sum_{i=0}^{2} t_{0i}(1+x_{0i})\rho^{1+i/3} + \frac{9}{160}t_1(1+x_1)\left(\frac{3\pi^2}{2}\right)^{2/3}\rho^{5/3}, \tag{2}$$

$$\frac{E_p^{01}}{A} = \frac{3}{16}\sum_{i=0}^{2} t_{0i}(1-x_{0i})\rho^{1+i/3} + \frac{9}{160}t_1(1-x_1)\left(\frac{3\pi^2}{2}\right)^{2/3}\rho^{5/3}, \tag{3}$$

$$\frac{E_p^{11}}{A} = \frac{27}{160}t_2(1+x_2)\left(\frac{3\pi^2}{2}\right)^{2/3}\rho^{5/3}. \tag{4}$$

where (t_i, x_i) are usual coefficients of the Skyrme functional, whereas (t_{0i}, x_{0i}) characterize the density-independent zero-range term and the two density-dependent ones.

Results are plotted against BHF predictions in Fig. 2. First, one can observe that the results are rather scattered. Second, the main source of binding, from the $(S,T) = (0,1)$ and $(1,0)$ channels, is not well described and the saturation mechanism is not captured. It is thus clear that, even though all four forces reproduce perfectly the PNM and SNM EOS, they do not have the same spin-isospin con-

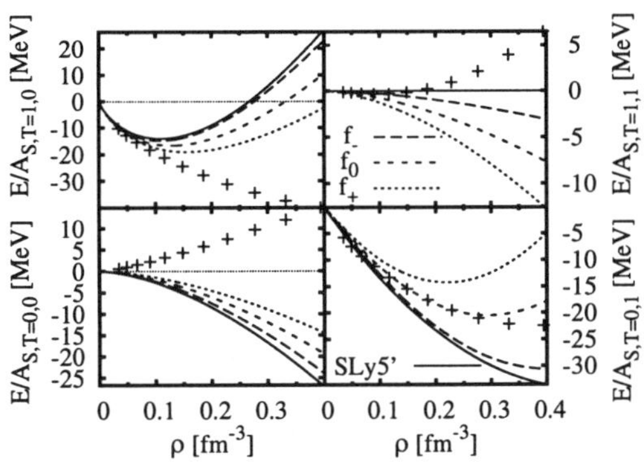

Figure 2. Energy per particle in each (S, T) channel for SNM, as a function of density. Crosses refer to the BHF calculations.[45]

tent, and that the latter is in general rather poor. Thus, fitting the global EOS is an important element but it does not mean that the functional has good spin-isospin properties. One needs to do more and fitting ab-initio predictions for the contribution per (S, T) channel seems to be a good idea. However, one needs to make sure that the data used are truly benchmarks calculations. This calls for predictions from other ab-initio methods.

The most obvious discrepancy appears in $(0, 0)$ and $(1, 1)$ channels where Skyrme and BHF data have opposite signs above saturation density. The SLy5 parameter set shows a particular behavior in the $(1, 1)$ channel due to the choice of $x_2 = -1$ to prevent ferromagnetic instabilities in PNM. Note that these two channels are taken care of in the Skyrme functional by the density-independent P-wave term only. The upper–right panel of Fig. 2 points out the tendency of Skyrme parameterizations to be attractive in polarized PNM and cause a collapse of its equation of state at high density. At lower densities, the BHF data show a distinctive behavior, being slightly attractive below ρ_0 and repulsive above. This feature cannot be matched by the standard Skyrme functional since it exhibits in this channel a monotonous behavior with density, whatever the value of (t_2, x_2).

It is also worth noticing that the failure in the $(1, 1)$ channel becomes more and more prominent as one makes the splitting of effective masses closer to the ab-initio predictions (force f4). The effective masses being governed by the momentum dependent terms of the force, it is not a surprise that the modification of the former impacts the $(0, 0)$ and $(1, 1)$ channels. What change in the coefficients entering Eqs. (1-4) stems only from the variation of m_v^* and the associated rearrangement of parameters in the functional, most notably the $C_{0,1}^{\Delta\rho}$ coefficients closely related to the surface and surface–symmetry energies. The quite tight requirements on the latter imply that the four parameters of the non–local terms in the standard Skyrme energy functional would be dramatically overconstrained if we were to add the (S,T)–channel decomposition in the fitting data.

The rather poor properties of the functional in the $(0,0)$ and $(1,1)$ channels, the degradation of the latter as the effective mass splitting is improved, the finite-size instability problem, the idea of using the ab-initio (S,T) contributions in the fit, call, at least, for a refinement of the odd–L term in the sense either of a density dependence or a higher–order derivative term. The latter being prone to numerical instabilities and interpretation problems, a density–dependent $\mathbf{k}' \cdot \mathbf{k}$ term remains as one of the next potential enhancements to be brought to the Skyrme energy functional. Phenomenological constraints on gradient terms are mainly related to the surface of nuclei, i.e. low–density regions. One can expect that, to first order, BHF data in $(S,T) = (1,1)$ channel can be matched with an extended functional while retaining a good agreement with other (experimental) data. It is less clear in the $(0,0)$ channel but further exploration of the extended parameter space may bring Skyrme and BHF data in better agreement.

4. Conclusions

Density Functional Theory faces important challenges in its quest for the universal energy density functional applicable over the entire nuclear chart with a true predictive power. In this contribution, we discuss some of those challenges. Then, we exemplify one of them by showing the use which can be made of further constraints from ab-initio calculations and how usual Skyrme functionals are too schematic/limited as well as over constrained as soon as one tries to bring more microscopic character to them.

Acknowledgments

The authors thank M. Bender and P.-H. Heenen for many valuable discussions. Above all, their thoughts go to the memory of Paul Bonche. This work was supported by the U.S. National Science Foundation under Grant No. PHY-0456903.

References

1. M. Bender, P.-H. Heenen and P.-G. Reinhard, *Rev. Mod. Phys.* **75**, p. 121 (2003).
2. J. P. Perdew, A. Ruzsinszky, J. Tao, V. N. Staroverov, E. Scuseria and G. I. Csonka, *Journ. Chem. Phys.* **123**, p. 062201 (2005).
3. W. Zuo, A. Lejeune, U. Lombardo and J. Mathiot, *Nucl. Phys.* **A706**, p. 418 (2002).
4. A. Akmal, V. R. Pandharipande and D. G. Ravenhall, *Phys. Rev.* **C58**, p. 1804 (1998).
5. S. C. Pieper and R. B. Wiringa, *Annu. Rev. Nucl. Part. Sci* **51**, p. 53 (2001).

6. P. Navratil and W. E. Ormand, *Phys. Rev.* **C68**, p. 034305 (2003).
7. D. J. Dean, J. R. Gour, G. Hagen, M. Hjorth-Jensen, K. Kowalski, T. Papenbrock, P. Piecuch and M. Wloch, *Nucl. Phys.* **A752**, p. 299 (2005).
8. G. F.Bertsch, B. Sabbey and M. Uusnakki, *Phys. Rev.* **C71**, p. 054311 (2005).
9. B. K. Agrawal, S. Shlomo and V. K. Au, *Phys. Rev.* **C72**, p. 014310 (2005).
10. M. Bender, P. Bonche, T. Duguet and P.-H. Heenen, *Phys. Rev.* **C69**, p. 064303 (2004).
11. V. Bernard and N. V. Giai, *Nucl. Phys.* **A348**, p. 75 (1980).
12. T. Otsuka, T. Suzuki, R. Fujimoto, H. Grawe and Y. Akaishi, *Phys. Rev. Lett.* **95**, p. 232502 (2005).
13. J. Dobaczewski, this volume, (2006).
14. T. Lesinski, M. Bender, K. Bennaceur, T. Duguet and J. Meyer, unpublished, (2006).
15. T. Otsuka, T. Matsuo and D. Abe, unpublished, (2006).
16. J. Dobaczewski and W. Nazarewicz, *Prog. Theor. Phys. Suppl.* **146**, p. 70 (2003).
17. T. Duguet, P. Bonche and P.-H. Heenen, *Nucl. Phys.* **A679**, p. 427 (2001).
18. Y. Yu and A. Bulgac, *Phys. Rev. Lett.* **90**, p. 222501 (2003).
19. A. Bulgac, *Phys. Rev.* **C65**, p. 051305 (2002).
20. T. Duguet, *Phys. Rev.* **C69**, p. 054317 (2004).
21. C. Shen, U. Lombardo and P. Schuck, *Phys. Rev.* **C71**, p. 054301 (2005).
22. F. Barranco, R. Broglia, G. Colo', E. Vigezzi and P. Bortignon, *Eur. Phys. J.* **A21**, p. 57 (2004).
23. T. Duguet, K. Bennaceur and P. Bonche, *YITP Report Series* **112**, B (2005).
24. H.-W. Hammer, *J. Phys.* **G31**, p. S1253 (2005).
25. S. K. Bogner, T. T. S. Kuo and A. Schwenk, *Phys. Rep.* **386**, p. 1 (2003).
26. S. K. Bogner, A. Schwenk, R. J. Furnstahl and A. Nogga, *Nucl. Phys.* **A763**, p. 59 (2005).
27. J. W. Negele and D. Vautherin, *Phys. Rev.* **C5**, p. 1472 (1972).
28. T. Duguet and P. Bonche, *Phys. Rev.* **C67**, p. 054308 (2003).
29. J. Dobaczewski, W. Nazarewicz, P. G. Reinhard and M. V. Stoitsov, unpublished, (2005).
30. M. Bender and T. Duguet, unpublished, (2006).
31. E. Chabanat, P. Bonche, P. Haensel, J. Meyer and R. Schaeffer, *Nucl. Phys.* **A627**, p. 710 (1997).
32. K. F. Liu and N. V. Giai, *Phys. Lett.* **B65**, p. 23 (1976).
33. P.-G. Reinhard, *Nucl. Phys.* **A649**, p. 305c (1999).
34. W. Zuo, I. Bombaci and U. Lombardo, *Phys. Rev.* **C60**, p. 024605 (2001).
35. Z.-Y. Ma, J. Rong, B.-Q. Chen, Z.-Y. Zhu and H.-Q. Song, *Phys. Lett.* **B604**, p. 170 (2004).
36. E. N. E. van Dalen, C. Fuchs and A. Faessler, *Phys. Rev.* **C72**, p. 065803 (2005).
37. M. Beiner, H. Flocard, N. V. Giai and P. Quentin, *Nucl. Phys.* **A238**, p. 29 (1975).
38. J. Bartel, P. Quentin, M. Brack, C. Guet and H.-B. Håkansson, *Nucl. Phys.* **A386**, p. 79 (1982).
39. F. Chappert, M. Girod and J.-F. Berger, private communication, (2006).
40. T. Lesinski, T. Duguet, K. Bennaceur and J. Meyer, unpublished, (2006).
41. B. Cochet, K. Bennaceur, P. Bonche, T. Duguet and J. Meyer, *Nucl. Phys.* **A731**, p. 34 (2004).
42. J. Dobaczewski, H. Flocard and J. Treiner, *Nucl. Phys.* **A422**, p. 103 (1984).

43. C. García-Recio, J. Navarro, N. V. Giai and N. N. Salcedo, *Ann. of Phys.* **214**, p. 293 (1992).
44. I. E. Lagaris and V. R. Pandharipande, *Nucl. Phys.* **A369**, p. 470 (1981).
45. M. Baldo, private communication, (2006).

BETA DECAY, THE R PROCESS, AND RIA

J. ENGEL
Department of Physics and Astronomy, University of North Carolina
Chapel Hill, North Carolina 27599-3255, USA
E-mail: engelj@physics.unc.edu

I review attempts to calculate beta-decay rates in neutron-rich nuclei created and destroyed in the r process, and discuss how we may do better. A facility for measuring the decay of neutron-rich nuclei will be valuable, even if it cannot reach all the nuclei along the r-process path.

1. Introduction

Several calculations of beta decay in r-process nuclei appeared shortly before the first Argonne/MSU/JINA/INT RIA workshop in January of 2004, and very few since then. Much of this article, therefore, resembles Ref. 1, which reviewed theoretical work on beta decay far from stability, though here I will be more explicit about ways to improve the calculations. First, some motivation:

1.1. *Beta-Decay Rates are Important*

Many of the nuclei made temporarily during r-process nucleosynthesis are currently inaccessible to experiment. Their properties, however, help determine the abundances of stable elements we observe in the solar system. Beta-decay rates are among the most important of these properties. Decay near the tops of closed-neutron-shell "ladders" in the r-process path is slow because the nuclei there are relatively close to the valley of stability. The small rates, which are partly responsible for large abundance peaks at $A \approx 80$, 130 and 195, determine the time it takes for nuclei to work themselves through the ladder. If we want to know how long the r process takes, we need good estimates of rates along the ladders (the largest of which, in the some of the nuclides closest to stability, have already been measured).

But even away from the ladders, beta-decay rates are important. They help determine, for example, the fraction of material that ends up outside (or inside) the three large abundance peaks. They also quantify the size of smaller abundance features, such as the "bump" in the rare-earth region. While measuring or calculating lifetimes of neutron-rich nuclei may not be necessary for determining the site of the r process, it is crucial for a detailed understanding of the r-process abundance curve.

1.2. *Calculating Beta Decay is Hard*

Unfortunately, beta-decay rates are not easy to calculate. They depend both on matrix elements and on phase space, which means that a nuclear model must provide good estimates of nuclear masses, excitation energies and matrix elements of decay operators.

The phase space associated with the decay to any particular state is roughly proportional to $(\Delta E)^{-5}$, so a modest error in the energy of a strongly populated state can be magnified in the decay rate. Most of the transition strength is governed by the Gamow-Teller operator $\vec{\sigma}\tau_-$, but in some nuclei forbidden transitions, associated with operators of the form $\vec{r}\vec{\sigma}\tau_-$, can compete. The allowed and forbidden operators are different and we must be able to calculate the matrix elements of both well.

One feature of neutron-rich nuclei makes the task easier: the further a nucleus is from stability, the larger its Q-value. This means that errors in excitation energies will be less significant than in nuclei near stability, where a small mistake can push one of the few accessible daughter states above threshold. When the Q-value is so large that a significant fraction of the total transition strength is below threshold, small errors in the distribution of that strength are much less significant.

2. Review of Approaches

Several theoretical schemes have been applied over the years to beta decay far from stability. I review several of the most prominent. Some methods emphasize global applicability, others self-consistency, and still others the comprehensive inclusion of nuclear correlations. None of the methods includes all the important correlations, however. As a result, the value of the axial vector coupling constant is renormalized from $g_A = 1.26$ to $g_A = 1.0$, at least for allowed transitions, in all the calculations discussed below.

2.1. *Macroscopic/Microscopic Mass Model + Schematic QRPA*

The essence of this and several related approaches discussed later is to divide the problem into two parts. First, the ground state masses of nuclei involved in the decay are calculated, then the excited states and transition matrix elements are generated. The macroscopic/microscopic mass model is based on the Strutinski method of adding shell-model effects to a collective description[2]. The latest refinement of one particular mass model is called the "Finite-Range Droplet Model" (FRDM)[3]. Daughter states accessible by Gamow-Teller decay are generated from the FRDM ground state through a separable interaction $V = 2\chi_{GT} : \vec{\sigma}\tau_{-} \cdot \vec{\sigma}\tau_{+} :$ (with $\chi_{GT} = 23$ MeV/A) in the charge-changing Quasiparticle Random Phase Approximation (QRPA). The latest version of this approach[4] also includes first-forbidden transitions in a statistical way, the result of which is to shorten important half-lives at $N = 82$ and (particularly) $N = 126$.

This model is not self-consistent — that is, the schematic interaction used in the QRPA is not related to the folded-Yukawa interaction used in the FRDM. Self consistency is desirable in principal, but only if the effective two-body interaction is well grounded. The advantage of the more phenomenological approach used here is that it can be more easily adjusted to data, and (because of the simple separable QRPA interaction) can be used in deformed nuclei as well as spherical ones, odd systems as well as even ones. More sophisticated approaches have not yet been consistently applied in deformed nuclei.

The average error cited in Ref. 4 for the lifetimes of nuclei living less than 1 second is about a factor of about 3. Although we really need to know the lifetimes of important r-process nuclei more accurately, we should bear in mind that these calculations are designed to reproduce all lifetimes, not just those important for the r process, and that for reasons noted above, the results are better than average in the r-process region far from stability.

Related global calculations with different prescriptions for obtaining the masses and mean fields exist (most notably a Nilsson-based approach[5] and the "Extended Thomas-Fermi with Strutinski Integral" (ETFSI) framework[6], which has been married to a less schematic QRPA) but have not been updated as recently as the calculations illustrated here.

2.2. *Self-Consistent Skyrme-HFB + QRPA*

In 1999 a paper appeared[7] that focused on the important "ladder" nuclei at closed neutron shells along the r-process path. These nuclei are spherical and therefore allowed a more sophisticated QRPA treatment than in the global approach discussed above. The calculation first employed the Hartree-Fock-Bogoliubov (HFB) approximation (in coordinate space because of the weak binding of neutron-rich nuclei) for even-even parents, then a pseudo-continuum[a] QRPA treatment of states in the daughters, in the "canonical basis" that diagonalizes the density matrix and

[a]Nuclei were enclosed in a hollow sphere of radius 20 fm

with the same Skyrme interaction as used in HFB. This self consistency makes the QRPA equivalent to the small-amplitude limit of time-dependent HFB, and allows systematic corrections.

The authors began by finding a Skyrme interaction[8] SKO$'$ that reproduced the energies and strengths of Gamow-Teller resonances reasonably well. They then adjusted a single parameter, the strength of $T = 0$ neutron-proton pairing, by fitting the half-lives of nuclei near those whose rates were being calculated. $T = 0$ pairing, neglected in the global calculations, played an important role in moving Gamow-Teller strength from the resonance down to low-lying excitations; without it the calculated half-lives would have been too long. [A weakness of these calculations was that the $T = 0$ pairing strength had to be different in each of the three regions of the nuclide chart that contained closed neutron shells.] The calculated transition rates for the r-process ladder nuclei — in the allowed approximation — were larger at $N = 50$ and 82 than those of the global approach by factors of 2 to 5. These factors become smaller, obviously, when forbidden transitions are added to the global calculations but are still larger than 1, and will grow again when the the two calculations include the same set of transitions. [At N=126 the HFB+QRPA rates were smaller than in the global calculations, but there were no measured nuclei to which to fit the $T = 0$ pairing strength there.]

How important was self consistency in these calculations? Though it is a necessary step towards a less phenomenological theory, the effective Skyrme interaction, which itself contains a lot of phenomenology, was far from perfect. The HFB+QRPA results are probably better than the more schematic ones discussed above, but the reason is more likely the limited focus on ladder nuclei than self consistency. Self consistency has proved more important in like-particle channels, where this model has been applied near the drip line with interesting results[9].

2.3. *Shell Model*

Shortly after the HFB calculations were published, a shell model calculation appeared for $N = 82$ r-process nuclei[10], supplemented by a later calculation[11] for $N = 126$. [A more recent calculation for $N = 82$, with similar results, has recently been published[12].] The shell model uses a smaller single-particle space than the QRPA, but includes many more correlations, some of which appear to be essential for an accurate description of low-lying strength. It is subject, however, to uncertainties in the effective interaction and operators, just like the other calculations. The shell-model rates at N=82 turned out to be even higher, by factors of two or more than those in the self-consistent HFB+QRPA calculations. They are also usually a little higher than those of both the HFB+QRPA and the global approach at $N = 126$. Forbidden transitions have yet to be included in the shell model, but doing so should be possible. Moving away from closed neutron shells, however, will be a formidable task, unlikely to be accomplished any time soon.

2.4. *Density-Functional + Finite-Fermi-Systems Theory*

Most recently, a density-functional/Greens-function-based version[13] of self-consistent HFB+QRPA (Density-Functional + Finite-Fermi-Systems (FFS) Theory — not quite self consistent but with a well-developed phenomenology) has been applied to spherical nuclei. The author was able to include forbidden transitions microscopically, something that the other methods have yet to do. Without forbidden transitions the rates are close to those of the HFB+QRPA, at least near $N = 82$. In that region the forbidden operators speed the transitions moderately, but at $N = 126$ they increase the rates by factors of several, so that the they are even faster than the those of the shell model (which included only allowed strength).

3. Development and Future of QRPA-like Approaches

The calculations described here are all better than those of the previous generation and all indicate that transitions far from stability are faster than previously believed, but don't fully agree with one another. Some calculations attempt to include forbidden transitions, others still do not. Only the FRDM + schematic QRPA can be applied in all nuclides. But the other methods can all be generalized and improved, and not just through obvious steps such as including forbidden decay. Here I discuss what has and should be done within the framework of self-consistent HFB+QRPA and FFS theory.

3.1. *Future of HFB+QRPA Beta Decay*

As noted above, self consistency is a virtue only in conjunction with a good effective interaction or energy functional. Shortly after the publication of HFB+QRPA beta decay rates, another paper[14] took the first steps toward improving Skyrme functionals by examining the effects of various "time-odd" terms (corresponding roughly to spin-dependent interactions) on Gamow-Teller strength distributions. The coefficients of these terms are usually not adjusted because they do not affect properties of even-even ground states. Their effect on states with nonzero spin, however, can be considerable. Though there were not enough data in spherical nuclei to determine all the time-odd parameters, by adjusting one (the Landau spin-isospin parameter g_0') the authors of Ref. 14 constructed an improved version of SkO′ that reproduces the strengths and energies of the available experimental resonances (see Fig. 1 below). That resulting interaction has since been applied to the calculation of real time-reversal violation[15] in heavy nuclei, which is sensitive to spin-dependent correlations. Once the HFB+QRPA scheme is generalized to deformed nuclei, more data can be examined and the time-odd part of the interaction further improved.

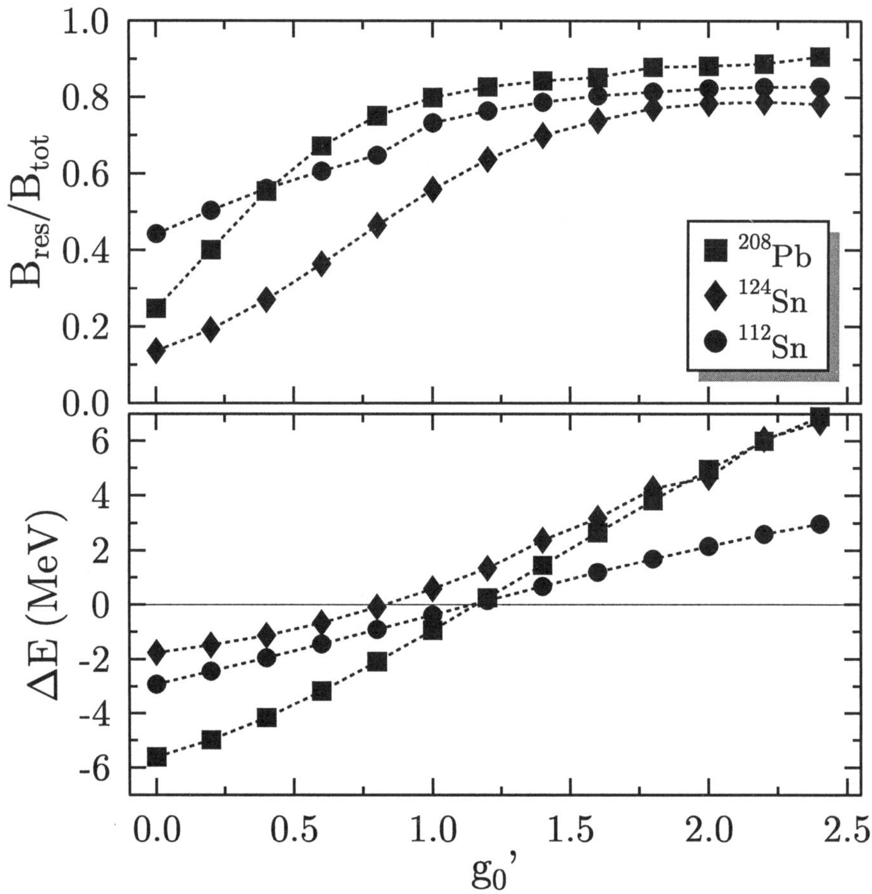

Figure 1. Error in Gamow-Teller resonance energy (bottom) and percentage of strength in the resonance (top) predicted by the Skyrme functional SkO′, as a function of the Landau parameter g_0' for three nuclei. The value $g_0' = 1.2$ adequately reproduces all the experimental data.

The extension to deformed nuclei is actually underway. A deformed HFB code already exists and has been applied throughout the table of isotopes[16]. The main reason deformed nuclei are harder in the QRPA is that the number of distinct single-particle configurations in a deformed potential is much larger than in a spherical potential, so that the basis of two-quasiparticle states can contain 10^5 – 10^6 states. Under such circumstance, for systematic calculations in many nuclei, one must resort to approximate diagonalization. Fortunately, similar problems in the shell model have led to the development and tuning of a powerful algorithm, the Lanczos method. A recent paper[17] develops a version of the algorithm that can be applied to the nonhermitian A and B matrices that make up the QRPA Hamiltonian, and tests the method with encouraging success in a model with a schematic interaction. I do not envision problems in adapting the method to the deformed-canonical-basis QRPA, either in the like-particle or charge-changing channels.

Finally, several groups have been working to improve the ordinary "time-even" parts of the Skyrme functional (see the paper, *e.g.* by J. Dobaczewski in these proceedings). The success of this program will have obvious benefits for Skyrme-QRPA calculations of beta decay.

3.2. *Future of FFS Beta Decay*

A lot of work has been done in recent years to extend FFS theory so that it goes beyond QRPA[18]. In the like-particle version, people are now doing calculations in which uncorrelated two-quasiparticle configurations are added to RPA phonons, producing strikingly different strength functions at low energies, where the presence of "pygmy" resonances has been a contentious issue for some time. An example, discussed in Ref. 19, is the E1 strength function in $^{40-44}$Ca. Low-energy strength is particularly important in beta decay because of the strong dependence of phase space on transition energy. The Green's-function formulation appears easier to augment, through approximations to self consistency, than the fully self-consistent HFB+QRPA, and although not enough personpower is involved to move to charge-changing transitions right away, the potential benefit to r-process calculations of doing so is clear.

4. Some Version of a Rare Isotope Accelerator Will be Useful

Although the machine formerly called RIA may not be constructed, even a good reacceleration facility would be of great use. It could measure the beta-decay rates of most nuclei along the r-process path, and would allow theory to calibrate itself well enough to handle those out of experimental reach. The value of nearby data to calibrate theory, even in "ab initio" calculations in light nuclei, cannot be overemphasized. Though a reacceleration facility may leave some r-process nuclear physics incompletely understood, its mass and lifetime measurements, combined with improving theory, should resolve most of the important issues surrounding beta decay.

Acknowledgments

This work was supported in part by the U.S. Department of Energy under grant DE–FG02–97ER41019.

References

1. J. Engel, in *The r-Process: The Astrophysical Origin of the Heavy Elements and Related Rare Isotope Accelerator Physics*, Y.Z. Qian, E. Rehm, H. Schatz, and F.-K. Thielemann, eds., (World Scientific, Singapore, 2004).
2. P. Ring. and R. Schuck *The Nuclear Many-Body Problem* (Springer-Verlag, Berlin, 1980).
3. P. Möller, J. R. Nix, W. D. Myers,and W. J. Swiatecki, *At. Data Nucl. Data Tables* **59**, 185 (1995).
4. P. Möller, B. Pfeiffer, and K. -L. Kratz, *Phys. Rev. C* **67**, 055802 (2003).
5. M. Homma et al., *Phys. Rev. C* **54**, 2972 (1999) and references therein.
6. I.N. Borzov and S. Goriely, *Phys. Rev. C* **62**, 035501 (2000) and references therein.
7. J. Engel, M. Bender, J. Dobaczewski, W. Nazarewicz, and R. Surman, *Phys. Rev. C* **60**, 014302 (1999).
8. P.-G. Reinhard *et al., Phys. Rev. C* **60**, 014316 (1999).
9. J. Terasaki, J. Engel, M. Bender, J. Dobaczewski, W. Nazarewicz, and M. Stoitsov, *Phys. Rev. C* **71**, 034310 (2005); J. Terasaki and J. Engel, `http://xxx.lanl.gov/nucl-th/0603005`.
10. G. Martinez-Pinedo and K. Langanke, *Phys. Rev. Lett.* **83**, 4502 (1999).
11. G. Martinez-Pinedo, *Nucl. Phys. A* **668**, 357c (2000).
12. B.A. Brown *et al., Nucl. Phys. A* **719**, 177c (2003).
13. I. N. Borzov, *Phys. Rev. C* **C67**, 025802 (2003).
14. M. Bender, J. Dobaczewski, J. Engel, and W. Nazarewicz, *Phys. Rev. C* **65**, 054322 (2002).
15. J. Dobaczewski and J. Engel, *Phys. Rev. Lett.* **94**, 232502 (2005); J.H de Jesus and J. Engel, *Phys. Rev. C* **72**, 045503 (2005).
16. M. V. Stoitsov, J. Dobaczewski, W. Nazarewicz, S. Pittel, and D.J. Dean, *Phys. Rev. C* **68**, 054312 (2003).
17. C.W. Johnson, G.F. Bertsch and W.D. Hazelton, *Comp. Phys. Comm.* **120**, 155 (1999).
18. S. Kamerdzhiev, J. Speth, and G. Tertychny, *Phys. Rep.* **393**, 1 (2004).
19. G. Tertychny, V. Tselyaev, S. Kamerdzhiev, F. Gruemmer, S. Krewald, J. Speth, A. Avdeenkov, and E. Litvinova, `http://xxx.lanl.gov/nucl-th/0603051`.

RIB MEASUREMENTS FOR STELLAR CORE COLLAPSE

W. R. HIX
Physics Division, Oak Ridge National Laboratory, Oak Ridge, TN 37831, USA
E-mail: raph@ornl.gov

O.E.B. MESSER
National Center for Computational Sciences, Oak Ridge National Laboratory, Oak Ridge, TN 37831, USA

G. MARTÍNEZ-PINEDO, K. LANGANKE
Gesellschaft für Schwerionenforschung, D-64291 Darmstadt, Germany

J. SAMPAIO
Centro de Fisica Nuclear da Universidade de Lisboa, 1649-003 Lisbon, Portugal

A. MEZZACAPPA, D. J. DEAN
Physics Division, Oak Ridge National Laboratory, Oak Ridge, TN 37831, USA

The most important nuclear interactions to the dynamics of stellar core collapse are electron capture and its inverse, neutrino capture. It has recently been re-discovered that electron captures on heavy nuclei (masses larger than 60) dominate electron captures on free protons. This dominance results in significant changes in the hydrodynamics of core collapse and bounce, compared to previous models where capture on free protons dominated. In this article we will review this recent work and discuss RIB measurements which may further our understanding of weak interactions with heavy nuclei and their importance to the mechanism of core collapse supernovae.

1. Introduction

Core collapse supernovae are among the most energetic events in the universe, emitting 10^{53} erg of energy, mostly in the form of neutrinos. Observationally categorized as Type II or Ib/c supernovae, these explosions mark the end of the life of a massive star and the formation of a neutron star or black hole. They play a preeminent role in the cosmic origin of the elements and serve as a principle heating mechanism for the interstellar medium.

The initiator of this explosion is the growth of the stellar iron core until it becomes too massive to be supported by electron degeneracy pressure. Electron capture therefore plays a large role in both the inception and rate of collapse. Bethe et al.[1] showed that the low entropy of the stellar core and resulting dominance of heavy nuclei over free nucleons causes electron capture on heavy nuclei to dominate the evolution of the electron fraction during the late stages of stellar evolution through the onset of stellar core collapse. Recent work[2,3] has shown that electron capture on nuclei plays a dominant role throughout collapse, significantly altering the electron fraction and entropy, thereby determining the strength and location of the initial supernova shock. These changes allow improvements in the treatment of electron capture to alter the initial conditions for the entire postbounce evolution of the supernova.

In the iron core, electron capture occurs predominantly via Gamow-Teller (GT) transitions changing protons in the $1f_{7/2}$ level of heavy nuclei into neutrons in the $1f_{5/2}$ level. The increasing density, and concomitant increase in the electron chemical potential, accelerates the capture of electrons on heavy nuclei and free protons in the core, producing electron neutrinos that initially escape, *deleptonizing* the core. The location at which the shock forms in the stellar core at bounce and the initial strength of the shock are largely set by the amount of deleptonization during collapse, rather than the preceding stellar evolution. Figure 1 summarizes the thermodynamic conditions throughout the core at bounce and displays the temperature, electron chemical potential (μ_e), and mean electron neutrino energy (E_{ν_e}) in MeV as functions of the matter density. Also shown is the electron fraction (Y_e) and representative nuclear mass (A). The kinks near $3 \times 10^7\,\mathrm{g\,cm^{-3}}$ mark the transition to the silicon shell. Deleptonization would be complete if electron capture continued without competition, but at densities between $10^{11-12}\,\mathrm{g\,cm^{-3}}$, the electron neutrinos become "trapped" in the core, reducing the rate of electron capture by Pauli blocking. Ultimately, the inverse neutrino capture reactions begin to compete with electron capture until the reactions are in weak equilibrium and net deleptonization of the core ceases on the core collapse time scale. The equilibration of electron neutrinos with matter occurs at densities between $10^{12-13}\,\mathrm{g\,cm^{-3}}$.

As the densities increase, the characteristic nuclei in the core increase in mass, owing to a competition between Coulomb contributions to the nuclear free energy and nuclear surface tension, until heavy nuclei are replaced by nuclear matter for mass densities near that of the nucleons in the nucleus ($\sim 10^{14}\,\mathrm{g\,cm^{-3}}$). For densities of order $10^{12}\,\mathrm{g\,cm^{-3}}$, nuclei with mass ~ 100 dominate. Fuller[4] realized that electron capture on heavy nuclei would soon be quenched in the Bethe et al.[1] picture as neutron numbers approach 40, filling the neutron $1f_{5/2}$ orbital.

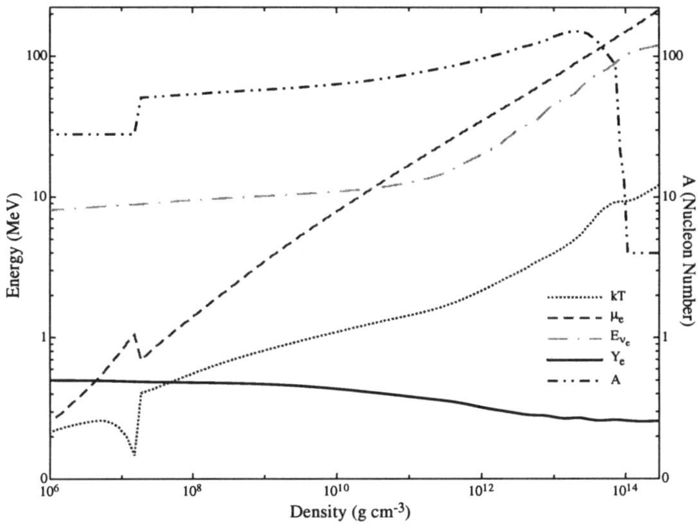

Figure 1. Typical energy scales as a function of density in the collapsed stellar core at bounce for a 25 $M_\odot$ progenitor. Shown are the thermal energy (kT), the electron chemical potential (μ_e) and the mean neutrino energy ($E_{\nu e}$). Also shown are the electron fraction (Y_e) and average mass of the heavy nuclei (A).

Independent particle model (IPM) calculations in the early eighties showed that neither thermal excitations nor forbidden transitions substantially alleviated this blocking,[4,5] leading to the belief that electron capture on protons dominated that on heavy nuclei during collapse. The dominance of captures on the small concentration of free protons results in a self-limiting response in the electron fraction. Unlike the concentration of heavy nuclei, changes of an order of magnitude in the free proton abundance can result from $\sim$ 10% changes in the electron fraction. Thus a reduction in the electron fraction inhibits further electron capture by greatly reducing the free proton fraction and therefore the rate of electron capture.[6] If electron capture on protons is the dominant process, this self-limitation during the course of collapse has been shown[7–9] to erase differences in electron fraction like those demonstrated in recent stellar evolution simulations.

2. Re-examination of the role of heavy nuclei

It is well known that the residual nuclear interaction (beyond the IPM) mixes the fp and gds shells, for example, making the closed $1g_{9/2}$ shell a magic number in stable nuclei ($N = 50$) rather than the closed fp shell ($N = 40$). This naturally calls into question whether the perception that electron capture on protons dominates during core collapse is an artifact of the IPM. To examine the possibility that nuclear electron capture is not quenched at $N = 40$, cross sections for charged-current electron and electron-neutrino capture on many nuclei up to at least $A \sim 100$ are needed to accurately simulate core deleptonization. Full shell model diagonalization calculations remain impossible in this regime due to the large number of available levels in the combined $fp + gds$ system.[10,11] Langanke et al.[12] developed a "hybrid" scheme, employing Shell Model Monte Carlo (SMMC) calculations of the temperature-dependent occupation of the single-particle orbitals to serve as input to Random Phase Approximation (RPA) calculations for allowed and forbidden transitions to calculate the capture rate. With this approach, Langanke et al.[2] (LMS) calculated electron capture rates for a sample of nuclei with $A = 66 - 112$.

Hix et al.[3] used these LMS rates, along with the shell model diagonalization rates of Langanke and Martínez-Pinedo[10] (LMP) for lighter nuclei, to develop a greatly improved treatment of nuclear electron capture. To calculate the needed abundances of the heavy nuclei, a Saha-like NSE was used, including Coulomb corrections to the nuclear binding energy,[13,14] but neglecting the effects of degenerate nucleons.[15] Comparison between the long standard Bruenn[16] prescription and this improved treatment of nuclear electron capture, which we will term the LMSH prescription, reveals two competing effects. In lower density regions, where the average nucleus is well below the $N = 40$ cutoff of electron capture on heavy nuclei, the Bruenn parameterization results in more electron capture than the LMSH case. This is similar to the reduction in the amount of electron capture seen in stellar evolution models[17] and thermonuclear supernova[18] models when the FFN rates are replaced by shell model calculations. In denser regions, the continuation of electron capture on heavy nuclei alongside electron capture on protons results in more electron capture in the LMSH case.

Figure 2 shows the impact of these new rates for a 25 solar mass progenitor (using the spherically symmetric general relativistic AGILE-BOLTZTRAN code[19–21]). Reductions at bounce of 9% in the central electron fraction, 12% in the central density and 3% in the central entropy are seen, as well as a 7% smaller velocity difference across the shock. The mass interior to the formation of the shock is reduced from .54 $M_{\odot}$to .49 $M_{\odot}$. A shift of this size is significant dynamically because the dissociation of .1 $M_{\odot}$ of heavy nuclei by the shock costs 10^{51} erg, the equivalent of the explosion energy. These effects are 50-75% of the size demonstrated by Hix et al.[3] for a 15 solar mass progenitor. Thus, across a wide range of supernova progenitor masses, the LMSH prescription results in the launch of a substantially weaker shock with more of the iron core overlying it, a change, in itself, that would inhibit a successful explosion.

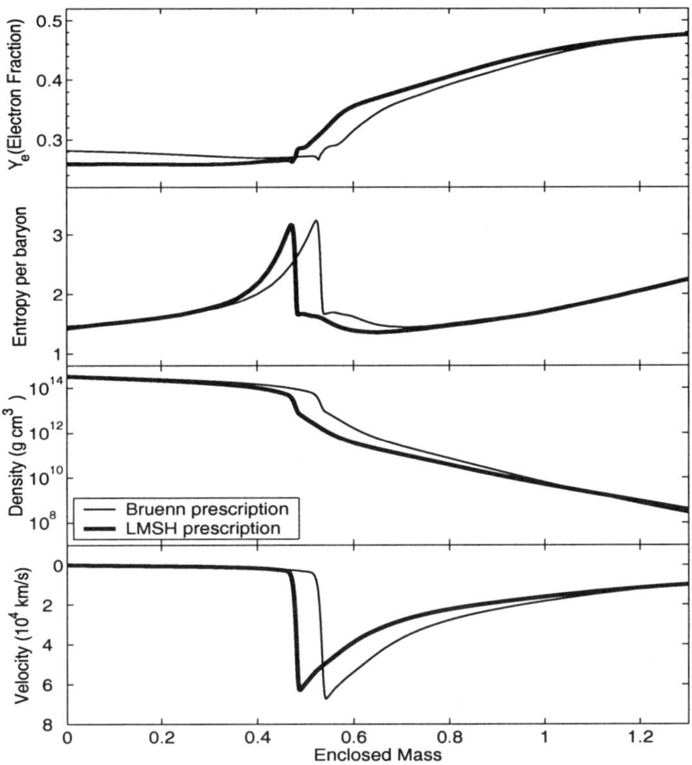

Figure 2. The electron fraction, entropy and velocity as functions of the enclosed mass at core bounce for a 25 $M_\odot$ model. The thin line is a simulation using the Bruenn parameterization while the thick line is for a simulation using the LMSH prescription. Both simulations implement General Relativity.

However, changes in the behavior of the outer layers also play an important role in the ultimate fate of the shock. The lesser neutronization in the outer layers (resulting primarily from the LMP rates) slows the collapse of these layers, which further diminishes the growth of the electron capture rate by reducing the rate at which the density increases. Reductions of a factor of 5 in density and 40% in velocity are found in the regions outside of the homologous core. Such changes reduce the ram pressure opposing the shock, easing its outward progress. In general relativistic, spherically symmetric models for a 15 $M_\odot$ progenitor,[17] these improvements allow the shock in the LMSH case to reach 168 km, relative to 166 km in the fiducial case. Thus the effect of lesser electron capture in the outer layers is comparable in size to the reduction of the mass of the homologous core. Additionally, changes in the electron capture rates also lead to changes in the core fluid gradients that may, in turn, drive fluid instabilities that are potentially important to the supernova mechanism. Within the inner 50 km, the entropy and lepton fraction gradients found in the LMSH 15 $M_{\odot}$ model are considerably different from those found in the fiducial model. Consequently, the more accurate treatment of electron capture may affect the multi-dimensional character of the supernova models by significantly altering the location, extent, and strength of proto-neutron star convection, or other potential fluid instabilities, in the core.

3. Rate Sensitivity

While the work of Hix et al.[3] and Langanke et al.[2] has established that core collapse is strongly affected when electron capture on heavy nuclei is not suppressed, this work does not address the sensitivity to the electron capture rates. One possibility is that the effects seen are due solely to breaking the self-limiting nature of electron capture on protons. Messer et al.[9] have however demonstrated considerable sensitivity in the models to uncertainties in the rates with a parameter study taking (for simplicity and reproducibility by other groups) the Bruenn prescription[16] as a starting point. In this prescription, the emissivity from heavy nuclei is proportional to the product of the number of protons in the $1f_{7/2}$ level and the number of neutron holes in the $1f_{5/2}$ level of the average nucleus, N_pN_h. It is the approach of this product to zero as $N \to 40$ that allows electron capture on protons to dominate in this prescription. Instead of letting the product N_pN_h vary as determined by the EOS, Messer et al.[9] set this product to several constant values in Newtonian collapse simulations. Figure 3 shows the effect of this variation on the velocity distribution at bounce, in comparison to the results of Newtonian models using the LMSH and Bruenn prescriptions. Clearly, a reduction in the total electron capture rate by a factor of 10 from those predicted by Langanke et al.[2] would erase the changes demonstrated by Hix et al.[3]. Likewise, a systematic increase by a factor of 10 would further reduce the initial PNS mass by at least 10%. Even global changes of a factor of 2, the confidence factor usually assigned to the better known shell model diagonalization rates, would have observable consequences, therefore further efforts to improve the treatment of electron capture on heavy nuclei are necessary.

4. The role of RIB measurements

While, the work summarized in §2 has reinvigorated interest in the role played by weak interactions with heavy nuclei in core collapse supernovae, the significant sensitivity in the models demonstrated in §3 at the level of the uncertainty in these rates calls for further improvements. The two distinct effects demonstrated, enhancement of electron capture in regions with densities larger than

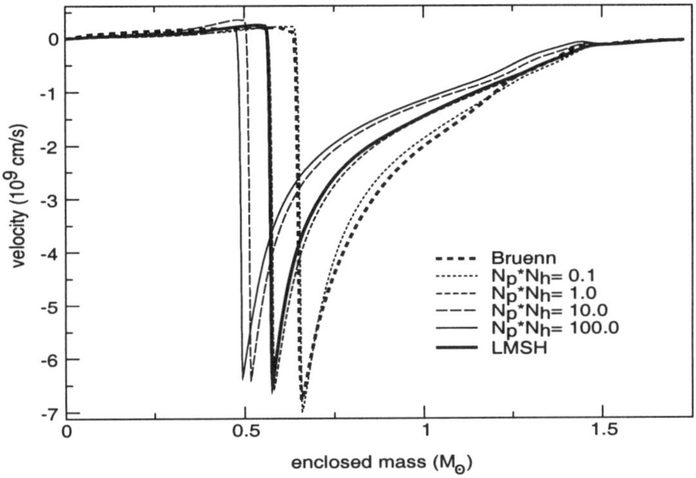

Figure 3. Comparison of the velocity structure of the core of a 15 $M_\odot$ star at bounce. The thick dotted and solid lines indicate models using the Bruenn and LMSH prescriptions, respectively. The thin lines show results of models using a modification of the Bruenn prescription where the product of number of protons in the $1f_{7/2}$ level and the number of neutron holes in the $1f_{5/2}$ level is held constant for the course of the simulation.

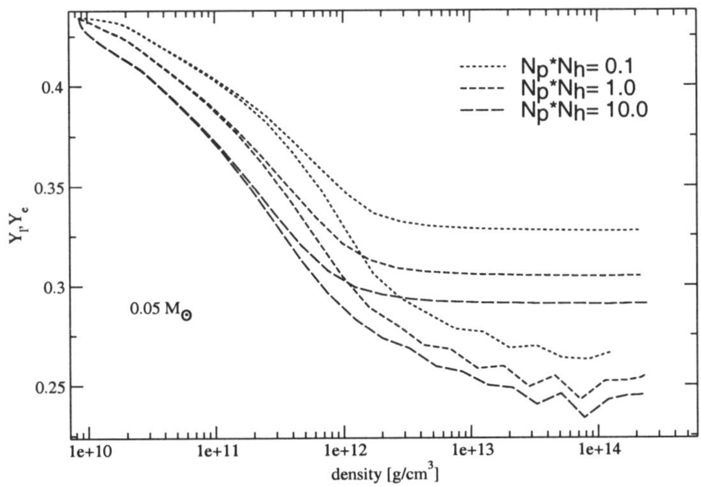

Figure 4. Evolution of the lepton (upper lines) and electron fraction (lower lines) during core collapse for an interior zone from three of the models shown in Fig. 3.

$3 \times 10^{10}\,\mathrm{g\,cm^{-3}}$, and reduced electron capture rates in lower density regions, are large, opposite, and, in spherical symmetric calculations, nearly equal in their

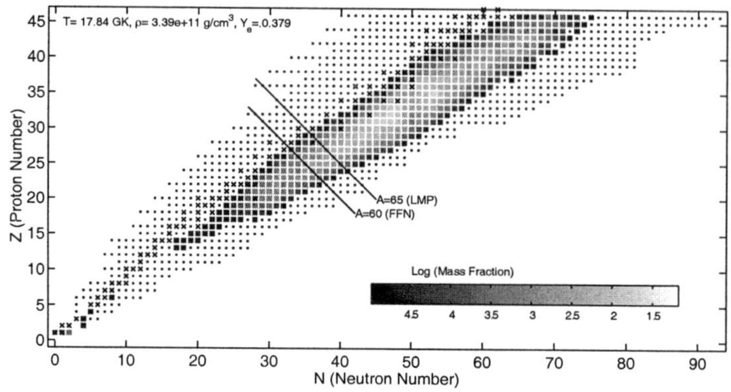

Figure 5. Details of the nuclear composition at a central density of 3×10^{11} $\mathrm{g\,cm^{-3}}$. The diagonal lines at A=60 and A=65 denoted the upper limits to the electron capture rate tabulations of FFN and LMP, respectively. For more massive nuclei, the LMSH prescription depends on the LMS rate tabulation.

aid/hinderance of a successful explosion. It is important to note that the rates which produce these two effects have separate origins, and therefore may change independently.

With densities of $\leq 10^{10}\,\mathrm{g\,cm^{-3}}$ and electron fractions $\sim .42 - .48$, the composition in the outer core is dominated by nuclei with $A \sim 65$. For many of these nuclei, shell model diagonalization calculations have been performed. Furthermore, since they lie along the neutron-rich edge of stability, measurements of the GT strength are also possible via charge exchange reactions as well as direct measurements of neutrino capture reaction rates. Comparison of measured GT strength distributions agree well with the theoretical calculations[22,23], implying that the rates are good (to within a factor of ~ 2) and thus far revealing no evidence for a systematic trend, which is far more important than random errors which would average out when the sum is taken over the nuclear distribution. This gives us confidence that the reduced electron capture and slower collapse of the outer layers described above is both qualitatively and quantitatively correct.

In the core, much heavier, more neutron-rich nuclei play an important role. Figure 4 shows that significant electron capture continues up to densities $\sim 10^{12}$ $\mathrm{g\,cm^{-3}}$, at which point the Pauli blocking significantly impedes electron capture and trapping levels out the lepton fraction. Changes in electron fraction beyond this point depend little on the electron capture rates. As can be seen in Figure 5, at such densities, a wide range of unstable nuclei as heavy as A=100 or more are present in significant abundance. At present, only the SMMC+RPA method used here and other, approximate methods[24] can provide electron capture rates for such nuclei. Experimental validation is extremely important, but will be difficult until next generation facilities like FAIR or RIA come online, since these nuclei are 4-6 units off of stability. We can have some confidence that the results summarized here are at least qualitatively correct because the electron chemical potentials of a couple tens of MeV(with kT $\sim 1 - 2\,\mathrm{MeV}$) found at these densities are larger than the typical Q-values, reducing the importance of the detailed nuclear structure. This places more emphasis on better determined quantities like the total GT strength rather than its distribution in energy, but experimental verification is important nonetheless.

Acknowledgements The authors thank M. Liebendörfer, S.W. Bruenn and H.-Th. Janka for fruitful discussions. The work has been partly supported by the U.S. Department of Energy, through the Scientic Discovery through Advanced Computing Program of the Office of Science and through the Advanced Simulations & Computing Academic Strategic Alliances Program Program (Grant No. B523820), by the U.S. National Science Foundation under contract PHY-0244783, by the Danish Research Council, by the Portugese Foundation for Science and Technology and by the Spanish MCyT and European Union ERDF under contracts AYA2002-04094-C03-02 and AYA200306128. This research used resources of the Center for Computational Sciences at Oak Ridge National Laboratory. Oak Ridge National Laboratory is managed by UT-Battelle, LLC, for the U.S. Department of Energy under contract DE-AC05-00OR22725.

References

1. H. A. Bethe, G. E. Brown, J. Applegate, and J. M. Lattimer, *Nucl. Phys. A* **324**, 487 (1979).
2. K. Langanke, G. Martinez-Pinedo, J. M. Sampaio, D. J. Dean, W. R. Hix, O. E. B. Messer, A. Mezzacappa, M. Liebendoerfer, H.-T. Janka, and M. Rampp, *Phys. Rev. Lett.* **90**, 241102 (2003), `astro-ph/0302459`.
3. W. R. Hix, O. E. B. Messer, A. Mezzacappa, M. Liebendoerfer, J. M. Sampaio, K. Langanke, D. J. Dean, and G. Martinez-Pinedo, *Phys. Rev. Lett.* **91**, 201102 (2003), `astro-ph/0310883`.
4. G. M. Fuller, *ApJ* **252**, 741 (1982).
5. J. Cooperstein and J. Wambach, *Nucl. Phys. A* **420**, 591 (1984).
6. F. D. Swesty, J. M. Lattimer, and E. S. Myra, *ApJ* **425**, 195 (1994).
7. M. Liebendörfer, O. E. B. Messer, A. Mezzacappa, W. R. Hix, F.-K. Thielemann, and K. Langanke, in *Proceedings of the 11th Workshop on Nuclear Astrophysics, Ringberg Castle, Tegernsee, Germany, February 11-16, 2002*, edited by W. Hillebrandt and E. Müller (2002), p. 126.
8. O. E. B. Messer, M. Liebendörfer, W. R. Hix, A. Mezzacappa, and S. W. Bruenn, in *Proc. of the ESO/MPA/MPE Workshop, From Twilight to Highlight: The Physics of Supernovae*, edited by W. Hillebrandt and B. Leibundgut (Heidelberg: Springer, 2003), p. 70.
9. O. Messer, W. Hix, A. Mezzacappa, and M. Liebendörfer, *ApJ* (2005).

10. K. Langanke and G. Martínez-Pinedo, *Nucl. Phys. A* **673**, 481 (2000).
11. G. S. Stoitcheva and D. J. Dean, in *Open Issues in Core Collapse Supernovae*, edited by G. Fuller and A. Mezzacappa (World Scientific, Singapore, 2005).
12. K. Langanke, E. Kolbe, and D. J. Dean, *Phys. Rev. C* **63**, 32801 (2001).
13. W. R. Hix and F.-K. Thielemann, *ApJ* **460**, 869 (1996).
14. E. Bravo and D. García-Senz, *MNRAS* **307**, 984 (1999).
15. M. F. El Eid and W. Hillebrandt, *A&AS* **42**, 215 (1980).
16. S. W. Bruenn, *ApJS* **58**, 771 (1985).
17. A. Heger, K. Langanke, G. Martínez-Pinedo, and S. E. Woosley, *Phys. Rev. Lett.* **86**, 1678 (2001).
18. F. Brachwitz, D. Dean, W. Hix, K. Iwamoto, K. Langanke, G. Martinez-Pinedo, K. Nomoto, M. R. Strayer, and F.-K. Thielemann, *ApJ* **536**, 934 (2000).
19. A. Mezzacappa and O. E. B. Messer, *J. Comp. Appl. Math* **109**, 281 (1999).
20. M. Liebendörfer, S. Rosswog, and F.-K. Thielemann, *ApJS* **141**, 229 (2002).
21. M. Liebendörfer, O. E. B. Messer, A. Mezzacappa, S. W. Bruenn, C. Y. Cardall, and F.-K. Thielemann, *ApJS* **150**, 263 (2004).
22. C. Bäumer, A. M. van den Berg, B. Davids, D. Frekers, D. de Frenne, E.-W. Grewe, P. Haefner, M. N. Harakeh, F. Hofmann, M. Hunyadi, et al., *Phys. Rev. C* **68**, 031303 (2003).
23. M. Hagemann, A. M. van den Berg, D. de Frenne, V. M. Hannen, M. N. Harakeh, J. Heyse, M. A. de Huu, E. Jacobs, K. Langanke, G. Martínez-Pinedo, et al., *Physics Letters B* **579**, 251 (2004).
24. J. Pruet and G. M. Fuller, *ApJS* **149**, 189 (2003).

SYMMETRY METHODS FOR EXOTIC NUCLEI

P. VAN ISACKER

GANIL, BP 55027, F-14076 Caen Cedex 5, France

A brief description is given of the use of algebraic methods in the context of the nuclear shell model and of the interacting boson model. As an example of such techniques, the problem of deuteron transfer between self-conjugate nuclei is treated in the context of a simplified interacting boson model which considers bosons without orbital angular momentum but with full spin–isospin structure.

1. Introduction

Nuclear models based on symmetry concepts can be separated rather naturally into two groups. In the first, a nucleus is considered as a system of interacting neutrons and protons, that is, fermions. This is the nuclear shell model with (residual) interactions, which is taken here as the starting point for the description of a nucleus. Two types of shell-model hamiltonian are discussed that can be solved analytically with algebraic techniques, namely those with pairing and those with quadrupole interactions. In the second group of symmetry-based models, a nucleus is treated as a system of interacting bosons that are of composite character and represent correlated pairs of nucleons. The choice of the different bosons is dictated by the nature of the nucleonic interactions and also depends on the particular nucleus under consideration. Connections can be established between the two classes of models.

We begin with a brief outline of the use of symmetry methods in the context of the shell model and the interacting boson model. As an illustration of how these methods can be applied to the physics of exotic nuclei, we discuss the problem of neutron–proton pairing in an algebraic framework.

2. The Nuclear Shell Model

A schematic shell-model hamiltonian which grasps the essential features of nuclei is of the form

$$H=\sum_{k=1}^{A}\left(\frac{p_k^2}{2m}+\frac{1}{2}m\omega^2r_k^2+\zeta\, l_k\cdot s_k+\kappa\, l_k\cdot l_k\right)+\sum_{k<l}^{A}V_2(\xi_k,\xi_l), \qquad (1)$$

where k, l run from 1 to A, the number of nucleons in the nucleus, ξ_k is a shorthand notation for the spatial coordinates, spin and isospin variables of nucleon k and m is the nucleon mass. The first term in this hamiltonian is of one-body type and includes the kinetic energy of the A nucleons as well as a simple representation of the mean field in terms of a harmonic-oscillator potential adjusted with spin–orbit and orbit–orbit terms. The second term in the hamiltonian is the residual two-body interaction which, in principle, depends in a complicated way on the mean-field potential and on the valence space available to the nucleons. In this sense it is an effective interaction. We neglect here interactions of higher order between the nucleons for which there is both theoretical and empirical evidence.

If the single-particle energy spacings are large in comparison with a typical matrix element of the residual interaction, nucleonic motion is independent and the shell model of independent particles results. The shell-model hamiltonian then has uncorrelated many-particle eigenstates which are Slater determinants constructed from the single-particle wave functions of the harmonic oscillator. If the residual interaction cannot be neglected, a genuine many-body problem is obtained which is much harder to solve. Interestingly, two types of residual interaction exist—pairing and quadrupole—which allow an analytic solution and have found fruitful application in nuclear physics.

2.1. *Pairing interaction*

Pairing models can be formulated either in LS or in jj coupling. The basic assumption is that the residual interaction among the valence nucleons has a pairing character. Thus, for example, in a single-j shell one considers an interaction which is attractive for two particles coupled to angular momentum $J = 0$ and zero otherwise. This is a reasonable, albeit schematic, approximation to the residual interaction between *identical* nucleons and hence can only be appropriate in semi-magic nuclei. The shell-model hamiltonian with a two-body pairing interaction can be diagonalized analytically in a space of identical fermions in a degenerate shell. This can be shown in a variety of ways but one elegant derivation relies on the existence of an SU(2) symmetry of the pairing hamiltonian, which is referred to as quasi-spin symmetry[1]. For even particle number n the ground state has a condensate structure of the form

$$(S_+)^{n/2} |\mathrm{o}\rangle, \tag{2}$$

where $|\mathrm{o}\rangle$ represents the vacuum (*i.e.* the doubly-magic core nucleus) and S_+ is a correlated two-fermion pair with orbital angular momentum $L = 0$ and spin $S = 0$. The conserved quantum number that emerges from these considerations is seniority[2], the number of nucleons not in pairs coupled to $L = 0$.

For a neutron and a proton there exists a different paired state with *parallel* spins. The orbital angular momentum of this pair again is $L = 0$ but now it has spin $S = 1$, and isospin $T = 0$ in order to guarantee overall anti-symmetry. Its total angular momentum is $J = 1$ and it can be referred to as a P pair. The most general pairing interaction for a system of neutrons and protons thus involves a *spin-singlet* and a *spin-triplet* term, and the pairing hamiltonian now involves two parameters, namely the strengths of the spin-singlet (or isovector) and spin-triplet (or isoscalar) interactions.

In general, this generalized pairing problem can only be solved numerically which, given the typical size of a nuclear shell-model space, can be a formidable task. For specific choices of the strengths, however, the solution can be obtained analytically[3]. The analysis reveals the existence of SO(8), which is the 'enveloping' algebra formed by the pair operators, their commutators, the commutators of these among themselves, and so on until a closed algebraic structure is obtained.

A special SO(8) solution occurs for the ground state of $N = Z$ nuclei. For example, in the limit of equal $T = 0$ and $T = 1$ pairing strengths, the exact ground-state solution can be written as[4]

$$(S_+ \cdot S_+ - P_+ \cdot P_+)^{n/4} |o\rangle. \tag{3}$$

This shows that the solution acquires a *quartet* structure in the sense that it reduces to a condensate of composite (approximate) bosons which each correspond to four nucleons. Since the boson in (3) is a scalar in spin and isospin, it can be thought of as an α particle. It can be shown[4] that a reasonable *ansatz* for the $N = Z$ ground-state wave function of a pairing hamiltonian with arbitrary isoscalar and isovector strengths is

$$(\cos\theta\, S_+ \cdot S_+ - \sin\theta\, P_+ \cdot P_+)^{n/4} |o\rangle, \tag{4}$$

where θ is a parameter which depends on the ratio of strengths.

The important (and as yet unanswered) question is now: To what extent does this quartet structure survive other terms that are present in a realistic shell-model hamiltonian, in particular, possible single-particle splittings? We shall return to this question in the discussion of the interacting boson model. But before doing so we briefly review the other class of solvable shell-model hamiltonians, namely those generated by the quadrupole interaction.

2.2. *Quadrupole interaction*

In the early days of nuclear physics, nuclei with a rotational-like spectrum were interpreted either with the droplet model of Bohr and Mottelson[5] or with the deformed single-particle shell model[6]. An understanding of rotational phenomena in terms of the spherical shell model, however, was lacking. Elliott's SU(3) model[7] did provide such an understanding from a symmetry perspective and has given rise to a class of solvable models adapted to deformed nuclei.

Since SU(3) is based on Wigner's supermultiplet model, a brief introduction to the latter is necessary. Wigner's supermultiplet or SU(4) model[8] assumes nuclear forces to be invariant under rotations in spin as well as isospin space. The physical relevance of Wigner's supermultiplet classification is connected with the short-range attractive nature of the residual interaction as a result of which states with spatial symmetry are favoured energetically. To see this point, consider an extreme form of a short-range interaction, namely a delta interaction. It has a vanishing matrix element in a spatially anti-symmetric two-nucleon state since in that case the wave function has zero probability of having the two particles at the same spot. In contrast, the matrix element is attractive in the spatially symmetric case. This result can be generalized to many nucleons, leading to the conclusion that the energy of a state depends on its SU(4) labels which then provide an approximate classification of states.

Wigner's supermultiplet model is a nuclear LS-coupling scheme and as such SU(4) is badly broken mainly by the spin–orbit interaction between nucleons. This break down of SU(4) symmetry increases with mass since the energy splitting of the spin doublets $l - 1/2$ and $l + 1/2$ increases with nucleon number A. In addition, SU(4) symmetry is also broken by the Coulomb interaction—an effect that also increases with A—and by other spin-dependent residual interactions.

In Wigner's supermultiplet model the spatial part of the wave function is left unspecified. It is only assumed that the total orbital angular momentum L is a good quantum number. The main feature of Elliott's model[7] is that it provides an orbital classification which incorporates rotational features. Elliott's model of rotation presupposes Wigner's SU(4) classification and assumes in addition that the residual interaction has a quadrupole character—a reasonable hypothesis if the valence shell contains neutrons and protons. The resulting shell-model hamiltonian can be analytically solved[9].

The situation can be summarized as follows. Elliott's SU(3) model provides a natural explanation of rotational phenomena, ubiquitous in nuclei, but it does so by assuming Wigner's SU(4) symmetry which is known to be badly broken in most nuclei. This puzzle has motivated much work since Elliott: How can rotational phenomena in nuclei be understood starting from a jj-coupling scheme induced by the spin–orbit term in the nuclear mean field? Arguably the most successful way to do so and to extend the applications of the SU(3) model to heavy nuclei is based upon the concept of pseudo-spin symmetry. This topic is treated in detail in the contribution of Jerry Draayer to this workshop. We shall not develop it further here but instead continue with the discussion of a shell-model based approximation method.

3. The Interacting Boson Model

As argued in the previous section, seniority-type as well as rotational-like spectra find a natural explanation in the nuclear shell model. A third, vibrational type of spectrum is frequently exhibited by nuclei, and its shell-model explanation is more problematic. Vibrational nuclei find an interpretation in terms of the geometric model of Bohr and Mottelson[5] where the vibrations are associated with (mainly quadrupole) oscillations of the nuclear surface. The disadvantage of this interpretation is that a transparent connection with the nuclear shell model is lacking. In this respect the interacting boson model (IBM) of Arima and Iachello[10] proved very useful. The model contains a vibrational and a rotational limit (as well as one which can be considered as intermediate), which connects well with the observed phenomenology of nuclei, and it can be brought into relation with the shell model.

In the original version of the IBM, applicable to even–even nuclei, the basic building blocks are s and d bosons[11]. The s and d bosons can be interpreted as correlated or Cooper pairs formed by two nucleons in the valence shell coupled to total angular momenta $J = 0$ and $J = 2$. This interpretation constitutes the basis of the connection between the IBM and the shell model[12].

Given the shell-model interpretation of the bosons, a low-lying collective state of an even–even nucleus with $2N$ valence nucleons is approximated as an N-boson state. Although the separate boson numbers n_s and n_d are not necessarily conserved, their sum $n_s + n_d = N$ is. This implies a total-boson-number conserving hamiltonian of the generic form

$$H = H_0 + H_1 + H_2 + H_3 + \cdots , \tag{5}$$

where the index refers to the order of the interaction in the generators of U(6). The first term is a constant that can be included to represent the nuclear binding energy of the core. The second term is the one-body part $H_1 = \epsilon_s n_s + \epsilon_d n_d$ where ϵ_s and ϵ_d are the single-boson energies of the s and d bosons. The third term in the hamiltonian represents the two-boson interaction and so on.

The characteristics of the most general hamiltonian which includes up to two-body interactions and its group-theoretical properties are by now well understood[13]. Numerical procedures exist to obtain the eigensolutions of a general IBM Hamiltonian but, as in the nuclear shell model, this many-body problem can be solved analytically for particular choices of boson energies and boson–boson interactions. For an IBM hamiltonian with up to two-body interactions between the bosons, three different analytical solutions or limits exist: the vibrational U(5)[14], the rotational SU(3)[15] and the γ-unstable SO(6) limit[16]. They are associated with the algebraic reductions

$$\mathrm{U}(6) \supset \left\{ \begin{array}{c} \mathrm{U}(5) \supset \mathrm{SO}(5) \\ \mathrm{SU}(3) \\ \mathrm{SO}(6) \supset \mathrm{SO}(5) \end{array} \right\} \supset \mathrm{SO}(3). \tag{6}$$

Where on the chart of nuclei can we expect such symmetries to occur? A simple way to obtain a quick answer to this question is based on plotting the ratio of the energies of the first 4^+ and 2^+ states (see Fig. 1). This ratio acquires

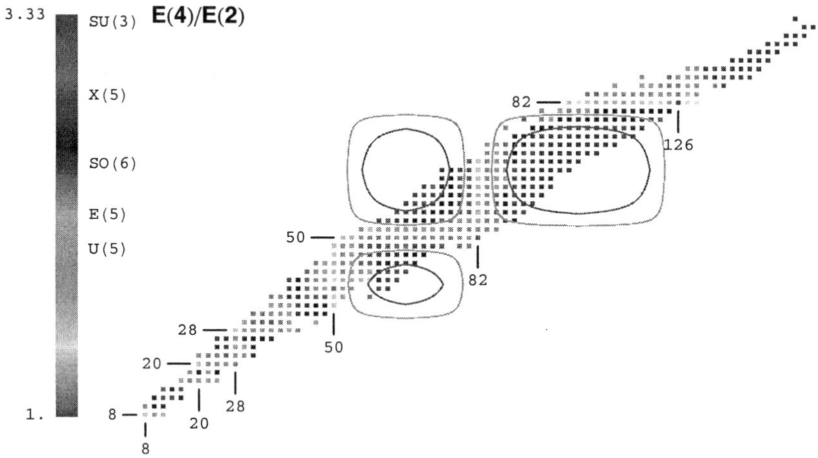

Figure 1. The ratio of excitation energies $E(4_1^+)/E(2_1^+)$ in all even–even nuclei with $N, Z \geq 8$ where it is known. The horizontal axis gives to the neutron number N and the vertical axis the proton number Z. The numbers 8, 20, 28, ... correspond to magic numbers of nucleons which stabilize the nucleus as a result of shell effects. The energy ratio is indicated by the colour coding shown on the left, together with the value it acquires in the different limits of the IBM and for the critical-point symmetries E(5) and X(5). Each pair of curves defines a locus of points where the product of valence neutron and proton numbers divided by their sum, $N_\mathrm{n}N_\mathrm{p}/(N_\mathrm{n}+N_\mathrm{p})$ has the value of 2.5 (outer) or 4.5 (inner). These are the values expected in nuclei transitional between U(5) and SO(6) [or E(5)] and between U(5) and SU(3) [or X(5)], respectively.

the values of 2, 2.5 and 3.33 in the U(5), SO(6) and SU(3) limits of the IBM and, furthermore, equals 2.19 and 2.91 in the critical-point symmetries E(5)[17] and X(5)[18], which can be considered as transitional between U(5) and SO(6), and U(5) and SU(3), respectively. On the other hand, on the basis of systematics one can associate[19] the ratio $E(4_1^+)/E(2_1^+)$ with the product of valence neutron and proton numbers divided by their sum, $P = N_\mathrm{n}N_\mathrm{p}/(N_\mathrm{n}+N_\mathrm{p})$. The curves in the figure correspond to values of P appropriate for transitional E(5) and X(5) nuclei. This then gives a rough idea of the possible occurrence of the IBM limits *if one assumes an unchanging shell structure in nuclei further removed from stability.* Any deviation from this expected structure could tentatively be considered as first evidence for changing shell structure in exotic nuclei.

The IBM has been used extensively to correlate energy spectra, electromagnetic decay and nucleon-transfer properties of many nuclei. In particular, it offers a natural framework to discuss the issue of two-nucleon transfer. Two-neutron and two-proton transfer have been analyzed in the early days of the model[20,21] using the neutron–proton version of the model, IBM-2[22], which includes neutron–neutron (nn) and proton–proton (pp) bosons. A description of deuteron transfer requires a more complicated version of the IBM with bosons corresponding to np pairs. Such extensions have been considered in the past and of particular relevance is the so-called IBM-4[23] since it contains np pairs with isospin $T = 0$ and $T = 1$. The full IBM-4 is a rich spectroscopic model[24] with bosons with orbital angular momentum $L = 0$ (s boson) or $L = 2$ (d boson), with intrinsic spin $S = 0$ or $S = 1$, and with isospin $T = 0$ (if $S = 1$) or $T = 1$ (if $S = 0$).

To avoid the complexity of the full IBM-4, it is instructive to confine the analysis to $L = 0$ bosons. This simplification preserves the complete spin–isospin structure of the model—crucial for the study of deuteron-transfer properties—and can be put to use in the analysis of the competition between isoscalar and isovector pairing in self-conjugate nuclei[25]. The dynamical algebra of the $L = 0$ IBM-4 is U(6), obtained from two vector bosons. One is vector in isospin while scalar in spin and the other boson is vector in spin while scalar in isospin.

The $L = 0$ IBM-4 can be considered as the boson equivalent of the SO(8) model discussed previously since its bosons have a spin–isospin structure identical to the pairs occurring in the fermion model. Furthermore, it can be shown[26] that the solvable limits of SO(8) have exact counterparts in the simplified IBM-4 which enables to establish a connection between the two models. This connection or mapping is valid not only for the pairing hamiltonian of SO(8) but also for less schematic hamiltonians which include, for example, single-particle splittings. The fermion-to-boson mapping then leads to the hamiltonian[27]

$$a\,C_2[\mathrm{SU}(4)] + b\,C_1[\mathrm{U}_S(3)] + c_1C_1[\mathrm{U}(6)] + c_2C_2[\mathrm{U}(6)] + d\,C_2[\mathrm{SO}_T(3)], \quad (7)$$

where $C_n[G]$ denotes a linear or quadratic ($n = 1,2$) Casimir operator of the algebra G. The first term in (7) is associated with SU(4) and implies equal single-boson energies and boson–boson interactions in the two isospin–spin channels $(T,S) = (0,1)$ and $(1,0)$. The second term breaks this equivalence between the two channels. The transition is governed by the single parameter b/a and intermediate results can be obtained by diagonalization[25].

The $L = 0$ IBM-4 hamiltonian (7) can be used to calculate nuclear binding energies[29] and in this way an estimate can be obtained of the parameters in (7). An example of results for nuclei in the first half of the 28–50 shell is shown in Fig. 2. The eigenstates of the hamiltonian (7) are entirely determined by the ratio

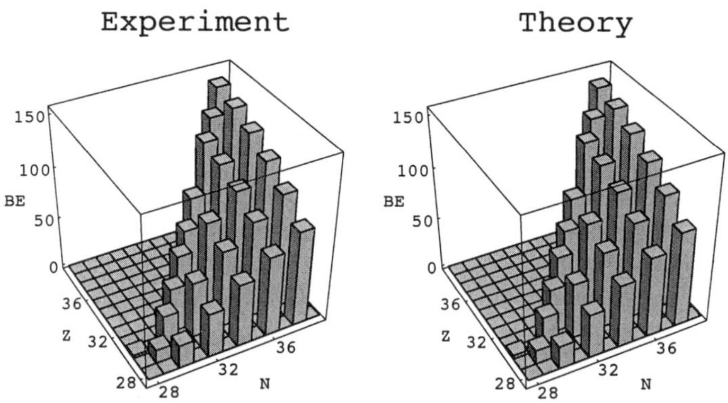

Figure 2. Binding energies (in MeV) of even–even (all N, Z) and odd–odd (only $N = Z$) nuclei in the first half of the 28–50 shell. All binding energies are relative to ^{56}Ni. The data are taken from Ref.[30]. The calculated values are obtained from (7) with $a = 0.238$, $b = 1.261$, $c_1 = -23.466$, $c_2 = -0.083$ and $d = 0.765$, in MeV. The root-mean-square deviation is 0.306 MeV.

b/a which comes out to be ~ 5 in the mass fit. This information can now be used to give an estimate of deuteron-transfer intensities.

Deuteron transfer is described in this model by the operators $s^\dagger$ ($p^\dagger$) for $T = 0$ ($T = 1$) transfer, where $s^\dagger$ ($p^\dagger$) creates a boson with $T = 0$ and $S = 1$ ($T = 1$ and $S = 0$), both with orbital angular momentum $L = 0$. From the matrix elements of these operators between the eigenstates of the hamiltonian (7) and from the mapping between the fermion SO(8) model and the $L = 0$ IBM-4, estimates can be obtained for the transfer intensities. We refer the reader to Ref.[27] for more details and simply give here the conclusion of the analysis. The favoured deuteron-transfer mode in this mass region has $T = 1$ rather than $T = 0$ character. Some appreciable strength of the latter can only be expected in the transfer from an even–even to an (excited) $T = 0$ state of an odd–odd nucleus.

4. Concluding Remark

This contribution has given a succinct overview of symmetry applications to nuclear structure in the context of the shell model and the interacting boson model. It is possible to present early nuclear structure models from a common perspective employing the language of symmetry and to demonstrate that they naturally lead into more recent developments such as the interacting boson model. By discussing more in detail the example of deuteron transfer in $N = Z$ nuclei, we hope to have convinced the reader that, even if these ideas of symmetry date back to the very beginnings of nuclear physics, they continue to inspire experiments at the forefront of today's research.

References

1. A.K. Kerman, *Ann. Phys. (NY)* **12**, 300 (1961).
2. G. Racah, *Phys. Rev.* **63**, 367 (1943).
3. B.H. Flowers and S. Szpikowski, *Proc. Phys. Soc.* **84**, 673 (1964).
4. J. Dobeš and S. Pittel, *Phys. Rev. C* **57**, 688 (1998).
5. A. Bohr and B.R. Mottelson, *Mat. Fys. Medd. Dan. Vid. Selsk.* **27**, No 16 (1953).
6. S.G. Nilsson, *Mat. Fys. Medd. Dan. Vid. Selsk.* **29**, No 16 (1955).
7. J.P. Elliott, *Proc. Roy. Soc. (London) A* **245**, 128 & 562 (1958).
8. E.P. Wigner, *Phys. Rev.* **51**, 106 (1937).
9. I. Talmi, *Simple Models of Complex Nuclei* (Harwood, Chur, 1993).
10. A. Arima and F. Iachello, *Phys. Rev. Lett.* **35**, 1069 (1975).
11. F. Iachello and A. Arima, *The Interacting Boson Model* (Cambridge University Press, Cambridge, 1987).
12. T. Otsuka, A. Arima and F. Iachello, *Nucl. Phys. A* **309**, 1 (1978).
13. O. Castaños, E. Chacón, A. Frank and M. Moshinsky, *J. Math. Phys.* **20**, 35 (1979).
14. A. Arima and F. Iachello, *Ann. Phys. (NY)* **99**, 253 (1976).
15. A. Arima and F. Iachello, *Ann. Phys. (NY)* **111**, 201 (1978).
16. A. Arima and F. Iachello, *Ann. Phys. (NY)* **123**, 468 (1979).
17. F. Iachello, *Phys. Rev. Lett.* **85**, 3580 (2000).
18. F. Iachello, *Phys. Rev. Lett.* **87**, 052502 (2001).
19. E.A. McCutchan, N.V. Zamfir and R.F. Casten, *J. Phys. G: Nucl. Part. Phys.* **31**, S1485 (2005).
20. A. Arima and F. Iachello, *Phys. Rev. C* **16**, 2085 (1977).
21. J.A. Cizewski, in *Interacting Bose-Fermi Systems in Nuclei*, ed. F. Iachello (Plenum, New York, 1981).
22. A. Arima, T. Otsuka, F. Iachello and I. Talmi, *Phys. Lett. B* **66**, 205 (1977).
23. J.P. Elliott and J.A. Evans, *Phys. Lett. B* **101**, 216 (1981).
24. P. Halse, J.P. Elliott and J.A. Evans, *Nucl. Phys. A* **417**, 301 (1984).
25. P. Van Isacker and D.D. Warner, *Phys. Rev. Lett.* **78**, 3266 (1997).
26. P. Van Isacker, J. Dukelsky, S. Pittel and O. Juillet, *J. Phys. G: Nucl. Part. Phys.* **24**, 1261 (1998).
27. P. Van Isacker, D.D. Warner and A. Frank, *Phys. Rev. Lett.* **94**, 162502 (2005).

28. O. Juillet and S. Josse, *Eur. Phys. J. A* **8**, 291 (2000).
29. E. Baldini-Neto, C.L. Lima and P. Van Isacker, *Phys. Rev. C* **65**, 064303 (2002).
30. G. Audi, A.H. Wapstra and C. Thibault, *Nucl. Phys. A* **729**, 337 (2003).

NEW DEVELOPMENTS IN NUCLEAR SUPERSYMMETRY: PICK-UP AND STRIPPING WITH SUSY AT RIA

R. BIJKER

ICN-UNAM, AP 70-543, 04510 Mexico DF, Mexico
E-mail: bijker@nucleares.unam.mx

In this contribution, I discuss the role of symmetries and algebraic methods in nuclear structure physics. In particular, I review some recent developments in nuclear supersymmetry and indicate possible applications for light nuclei in the sd- and pf-shell.

1. Introduction

In recent years, nuclear structure has seen an impressive progress in the development of *ab initio* methods (no-core shell model, Green's Function Monte Carlo, Coupled Clusters, ...), mean-field techniques and effective field theories for which the ultimate goal is *an exact treatment of nuclei utilizing the fundamental interactions between nucleons* [1]. All involve large scale calculations and therefore relie heavily on the available computing power and the development of efficient algorithms to obtain the desired results.

A different approach is that of symmetries and algebraic methods. Rather than trying to solve the complex nuclear many-body problem numerically, one tries to identify effective degrees of freedom, effective or dynamical symmetries, etc. Aside from their esthetic appeal, symmetries provide energy formula, selection rules and closed expressions for electromagnetic transition rates and transfer strengths which can be used as benchmarks to study and interpret the experimental data, even if these symmetries may be valid only approximately. Symmetries have played an important role in the history of nuclear physics. Examples are isospin symmetry, the Wigner supermultiplet theory, special solutions to the Bohr Hamiltonian, the Elliott model, pseudo-spin symmetries and the dynamical symmetries of the IBM and its extensions.

The aim of this contribution is to discuss the role of symmetries and algebraic methods in nuclear structure physics. In particular, I review some recent results obtained for nuclear supersymmetry in the Pt-Au region and discuss some implications for light nuclei in the sd- and pf-shell.

2. Dynamical supersymmetries in nuclear physics

Dynamical supersymmetries were introduced [2] in nuclear physics in the context of the Interacting Boson Model (IBM) and its extensions. The IBM describes collective excitations in even-even nuclei in terms of a system of interacting monopole ($s^\dagger$) and quadrupole ($d^\dagger$) bosons, which altogether can be denoted by $b_i^\dagger$ with angular momentum $l = 0, 2$. The bosons are associated with the number of correlated proton and neutron pairs, and hence the number of bosons N is half the number of valence nucleons [3].

In general, the IBM Hamiltonian has to be diagonalized numerically to obtain the energy eigenvalues and wave functions. There exist, however, special situations in which the eigenvalues can be obtained in closed, analytic form. These situations correspond to dynamical symmetries which arise, whenever the Hamiltonian is expressed in terms of Casimir invariants of a chain of subgroups of $G = U(6)$: the $U(5)$ limit for vibrational nuclei, the $SU(3)$ limit for rotational nuclei and the $SO(6)$ limit for γ-unstable nuclei [3]. For each one of the dynamical symmetries a set of closed analytic expressions has been derived for energies, electromagnetic transitions, quadrupole moments and other observables of interest which can be used to classify and interpret the available experimental data in a qualitative way.

For odd-mass nuclei the IBM has been extended to include single-particle degrees of freedom [4]. The ensuing Interacting Boson-Fermion Model (IBFM) has as its building blocks N bosons with $l = 0, 2$ and $M = 1$ fermion ($a_j^\dagger$) with $j = j_1, j_2, \ldots$ [5]. The IBM and IBFM can be unified into a supersymmetry (SUSY) [6]

$$U(6/\Omega) \supset U(6) \otimes U(\Omega) \,, \tag{1}$$

where $\Omega = \sum_j (2j+1)$ is the dimension the fermion space. In this framework, even-even and odd-even nuclei form the members of a supermultiplet which is characterized by $\aleph = N + M$, *i.e.* the total number of bosons and fermions. Supersymmetry distinguishes itself from other symmetries in that it includes, in addition to transformations among fermions and among bosons, also transformations that change a boson into a fermion and *vice versa* (see Table 1).

Model	Generators	Invariant	Symmetry
IBM	$b_i^\dagger b_j$	N	$U(6)$
IBFM	$b_i^\dagger b_j \,,\ a_k^\dagger a_l$	N, M	$U(6) \otimes U(\Omega)$
SUSY	$b_i^\dagger b_j \,,\ a_k^\dagger a_l \,,\ b_i^\dagger a_k \,,\ a_k^\dagger b_i$	$\aleph = N + M$	$U(6/\Omega)$

3. Supersymmetry in heavy nuclei

Dynamical nuclear supersymmetries correspond to very special forms of the Hamiltonian which may not be applicable to all regions of the nuclear chart, but nevertheless many nuclei have been found to provide experimental evidence for supersymmetries in nuclei [5,7]. Especially, the mass region $A \sim 190$ has been a rich source of empirical evidence for the existence of (super)symmetries in nuclei. The even-even nucleus ^{196}Pt is the standard example of the $SO(6)$ limit of the IBM [8]. The odd-proton nuclei ^{191,193}Ir and ^{193,195}Au were suggested as examples of the $Spin(6)$ limit [2], in which the odd proton is allowed to occupy the $2d_{3/2}$ orbit of the 50-82 proton shell, whereas the pairs of nuclei ^{190}Os - ^{191}Ir, ^{192}Os - ^{193}Ir, ^{192}Pt - ^{193}Au and ^{194}Pt - ^{195}Au have been analyzed as examples of a $U(6/4)$ supersymmetry [6].

The odd-neutron nucleus ^{195}Pt, together with ^{194}Pt, were studied in terms of a $U(6/12)$ supersymmetry [9], in which the odd neutron occupies the $3p_{1/2}$, $3p_{3/2}$ and $2f_{5/2}$ orbits of the 82-126 neutron shell. In this case, the neutron angular momenta are decoupled into a pseudo-orbital part with $\tilde{l} = 0, 2$ and a pseudo-spin part with $\tilde{s} = \frac{1}{2}$. This supersymmetry scheme arises from the equivalence between the values of the angular momenta of the pseudo-orbital part and those of the bosons of the IBM.

The concept of nuclear SUSY was extended in 1985 to include the neutron-proton degree of freedom [10]. In this case, a supermultiplet consists of an even-even, an odd-proton, an odd-neutron and an odd-odd nucleus. Spectroscopic studies of heavy odd-odd nuclei are very difficult due the high density of states. Almost 15 years after the prediction of the spectrum of the odd-odd nucleus by nuclear supersymmetry, it was shown experimentally that the observed spectrum of the nucleus ^{196}Au is amazingly close to the theoretical one [11]. At present, the best experimental evidence of a supersymmetric quartet is provided by the ^{194,195}Pt and ^{195,196}Au nuclei as an example of the $U(6/12)_\nu \otimes U(6/4)_\pi$ supersymmetry. This supermultiplet is characterized by $\aleph_\pi = 2$ and $\aleph_\nu = 5$.

In this case, the excitation spectra of the supersymmetric quartet of Pt and Au nuclei are described simultaneously by the energy formula

$$\begin{aligned} E = {} & \alpha\,[N_1(N_1+5) + N_2(N_2+3) + N_1(N_1+1)] \\ & + \beta\left[\Sigma_1(\Sigma_1+4) + \Sigma_2(\Sigma_2+2) + \Sigma_3^2\right] \\ & + \gamma\left[\sigma_1(\sigma_1+4) + \sigma_2(\sigma_2+2) + \sigma_3^2\right] \\ & + \delta\left[\tau_1(\tau_1+3) + \tau_2(\tau_2+1)\right] + \epsilon\, J(J+1) + \eta\, L(L+1)\,. \end{aligned} \tag{2}$$

The coefficients α, β, γ, δ, ϵ and η have been determined in a simultaneous fit of the excitation energies of the nuclei ^{194,195}Pt and ^{195,196}Au [12]. Fig. 1 shows the

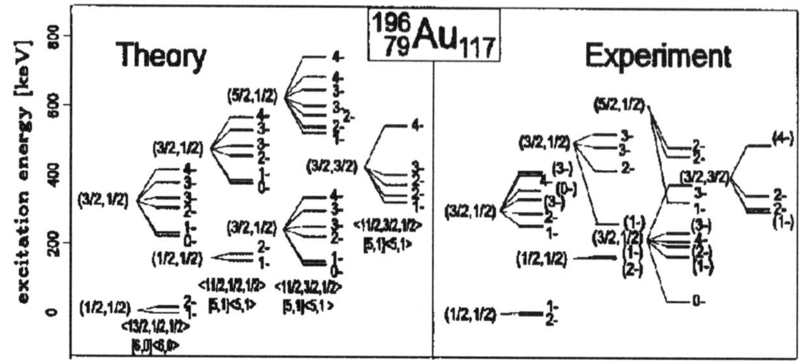

Figure 1. Comparison between the energy spectrum of the negative parity levels in the odd-odd nucleus ^{196}Au and that obtained for the $U(6/12)_\nu \otimes U(6/4)_\pi$ supersymmetry [12].

results for the odd-odd nucleus ^{196}Au.

3.1. *One-nucleon transfer reactions*

The supersymmetric quartet of nuclei is described by a single Hamiltonian, and hence the wave functions, transition and transfer rates are strongly correlated. As an example of these correlations, I consider here the case of one-proton transfer reactions between the Pt and Au nuclei. One-proton transfer reactions between different members of the same supermultiplet provide an important test of supersymmetries, since it involves the transformation of a boson into a fermion or *vice versa*, but it conserves the total number of bosons plus fermions.

The operators that describe one-proton transfer reactions in the $U(6/12)_\nu \otimes U(6/4)_\pi$ supersymmetry are, in lowest order, given by

$$P_1^\dagger = \alpha_1 \left[-\sqrt{\frac{1}{6}} \left(\tilde{s}_\pi \times a^\dagger_{\pi,3/2} \right)^{(3/2)} + \sqrt{\frac{5}{6}} \left(\tilde{d}_\pi \times a^\dagger_{\pi,3/2} \right)^{(3/2)} \right] ,$$

$$P_2^\dagger = \alpha_2 \left[\sqrt{\frac{5}{6}} \left(\tilde{s}_\pi \times a^\dagger_{\pi,3/2} \right)^{(3/2)} + \sqrt{\frac{1}{6}} \left(\tilde{d}_\pi \times a^\dagger_{\pi,3/2} \right)^{(3/2)} \right] . \qquad (3)$$

As a consequence of the selection rules, the operator $P_1^\dagger$ only excites the ground state of the Au nuclei, whereas $P_2^\dagger$ populates, in addition to the ground state, also an excited state [13]. The ratio of the intensities of one-proton transfer to the excited state and to the ground state $R = I(\text{gs} \rightarrow \text{exc})/I(\text{gs} \rightarrow \text{gs})$ does not depend on any parameter, and is given by

$$R_1(^{195}\text{Pt} \rightarrow ^{196}\text{Au}) = R_1(^{194}\text{Pt} \rightarrow ^{195}\text{Au}) = 0 \,,$$
$$R_2(^{195}\text{Pt} \rightarrow ^{196}\text{Au}) = R_2(^{194}\text{Pt} \rightarrow ^{195}\text{Au}) = \frac{9(N+1)(N+5)}{4(N+6)^2} \,, \qquad (4)$$

for $P_1^\dagger$ and $P_2^\dagger$, respectively. The available experimental data from the proton stripping reactions $^{194}\text{Pt}(\alpha, t)^{195}\text{Au}$ and $^{194}\text{Pt}(^3\text{He}, d)^{195}\text{Au}$ [14] shows that the $J = 3/2$ ground state of ^{195}Au is excited strongly. The relatively small strength to excited $J = 3/2$ states suggests that the operator $P_1^\dagger$ of Eq. (3) be used to describe the experimental data.

The equality of the ratios of the one-proton transfer reactions $^{194}\text{Pt} \rightarrow {}^{195}\text{Au}$ and $^{195}\text{Pt} \rightarrow {}^{196}\text{Au}$ is a direct consequence of the supersymmetry classification. This prediction has been tested experimentally using the (^{3}He, d) reaction on ^{194}Pt and ^{195}Pt targets. The results are being analyzed at the moment [15].

In addition, in a supersymmetry scheme it is possible to establish explicit relations between the intensities of these two transfer reactions, *i.e.* the one-proton transfer reaction intensities between the (ground state of the) Pt and Au nuclei are related by

$$I(^{195}\text{Pt} \rightarrow {}^{196}\text{Au}) = \frac{2L+1}{4}\, I(^{194}\text{Pt} \rightarrow {}^{195}\text{Au}) \,. \qquad (5)$$

This correlation can be derived in a general way only using the symmetry relations that exist between the wave functions of the even-even, odd-proton, odd-neutron and odd-odd nuclei of a supersymmetric quartet. It is important to point out, that Eqs. (4 and (5) are parameter-independent predictions which are a direct consequence of nuclear SUSY and which can be tested experimentally.

3.2. *Two-nucleon transfer reactions*

Two-nucleon transfer reactions probe the structure of the final nucleus through the exploration of two-nucleon correlations that may be present. The spectroscopic strengths not only depend on the similarity between the states in the initial and final nucleus, but also on the correlation of the transferred pair of nucleons.

In this section, the recent data on the $^{198}\text{Hg}(\vec{d}, \alpha)^{196}\text{Au}$ reaction [16] are compared with the predictions from the $U_\nu(6/12) \otimes U_\pi(6/4)$ supersymmetry. This reaction involves the transfer of a proton-neutron pair, and hence measures the neutron-proton correlation in the odd-odd nucleus. The spectroscopic strengths G_{LJ}

$$G_{LJ} = |\sum_{j_\nu j_\pi} g^{LJ}_{j_\nu j_\pi} \langle ^{196}\text{Au} \| \, (a^\dagger_{j_\nu} a^\dagger_{j_\pi})^{(\lambda)} \, \| ^{198}\text{Hg} \rangle \,|^2 \,, \qquad (6)$$

depend on the reaction mechanism via the coefficients $g^{LJ}_{j_\nu j_\pi}$ and on the nuclear structure part via the reduced matrix elements.

In order to compare with experimental data we calculate the relative strengths $R_{LJ} = G_{LJ}/G^{\rm ref}_{LJ}$, where $G^{\rm ref}_{LJ}$ is the spectroscopic strength of the reference state. Fig. 2 shows the experimental and calculated ratios R_{LJ}. The reference states are easily identified since they are normalized to one.

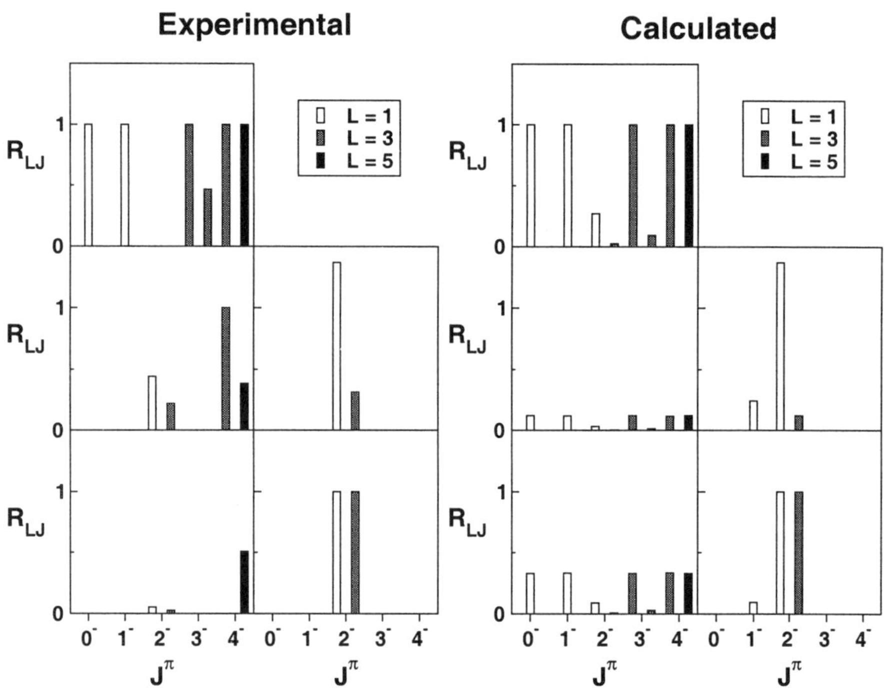

Figure 2. Ratios of spectroscopic strengths. The two columns in each frame correspond to states with $Spin(5)$ labels $(\tau_1, \tau_2) = (\frac{3}{2}, \frac{1}{2})$ and $(\frac{1}{2}, \frac{1}{2})$, respectively. The rows are characterized by the labels $[N_1, N_2]$, $(\Sigma_1, \Sigma_2, 0)$, $(\sigma_1, \sigma_2, \sigma_3)$. From bottom to top we have (i) $[6, 0]$, $(6, 0, 0)$, $(\frac{13}{2}, \frac{1}{2}, \frac{1}{2})$, (ii) $[5, 1]$, $(5, 1, 0)$, $(\frac{11}{2}, \frac{1}{2}, -\frac{1}{2})$ and (iii) $[5, 1]$, $(5, 1, 0)$, $(\frac{11}{2}, \frac{3}{2}, \frac{1}{2})$.

The ratios of spectroscopic strengths to final states with $(\tau_1, \tau_2) = (\frac{3}{2}, \frac{1}{2})$ provide a direct test of the nuclear wave functions, since they only depend on the nuclear structure part [17]

$$R_{2,LJ} = \frac{N+4}{15N} ,$$
$$R_{3,LJ} = \frac{2(N+4)(N+6)}{15N(N+3)} , \tag{7}$$

for different final states corresponding to the left panel of the second and third row in Fig. 2. The numerical values are 0.12 and 0.33, respectively ($N = 5$). In general, there is good overall agreement between the experimental and theoretical values, especially if we take into account the simple form of the operator in the calculation of the two-nucleon transfer reaction intensities.

4. Nuclear supersymmetry in light nuclei

Dynamical supersymmetries correspond to special solutions of the Hamiltonian. They occur whenever the even-even nucleus can be described by one of the dynamical symmetries of the IBM, and the odd nucleon occupies specific single-particle orbits which lie close to the Fermi surface [18].

In light nuclei, examples may be found in the sd-shell for which the orbital angular momenta of the $2s_{1/2}$, $1d_{3/2}$ and $1d_{5/2}$ orbits match the angular momenta of the bosons [19]. Another area of interest may be the beginning of the pf-shell where the valence nucleons occupy the orbits $2p_{1/2}$, $1p_{3/2}$ and $1f_{5/2}$ which can be treated in a pseudo-spin coupling scheme as $\tilde{l} = 0, 2$ and $\tilde{s} = \frac{1}{2}$.

In heavy nuclei, where the active protons and neutrons occupy different major shells, the s- and d-bosons are associated with correlated pairs of identical nucleons with isospin $T = 1$ and $M_T = \pm 1$. For light nuclei the situation is different, since the valence protons and neutrons occupy the same major shell. This observation has led to the introduction of an isospin invariant IBM in which the s- and d-bosons can have spin and isospin $(S,T) = (0,1)$ and $(1,0)$ [20] which leads to the algebraic structure

$$U^B(36) \supset U^B_L(6) \otimes SU^B_{ST}(6) \supset U^B_L(6) \otimes SU^B_{ST}(4) . \tag{8}$$

The subscripts refer to the orbital part (L) and the spin-isospin part (ST). The group structure of the odd nucleon in the sd-shell and in the beginning of the pf-shell is the same and is given by

$$U^F(24) \supset U^F_L(6) \otimes SU^F_{ST}(4) . \tag{9}$$

Whereas for the sd-shell, $SU_{ST}(4)$ refers to the Wigner supermultiplet $SU(4)$ symmetry, for the pf-shell it represents the pseudo-$SU(4)$ symmetry which follows from the combined invariance in pseudo-spin and isospin [21]. It has been

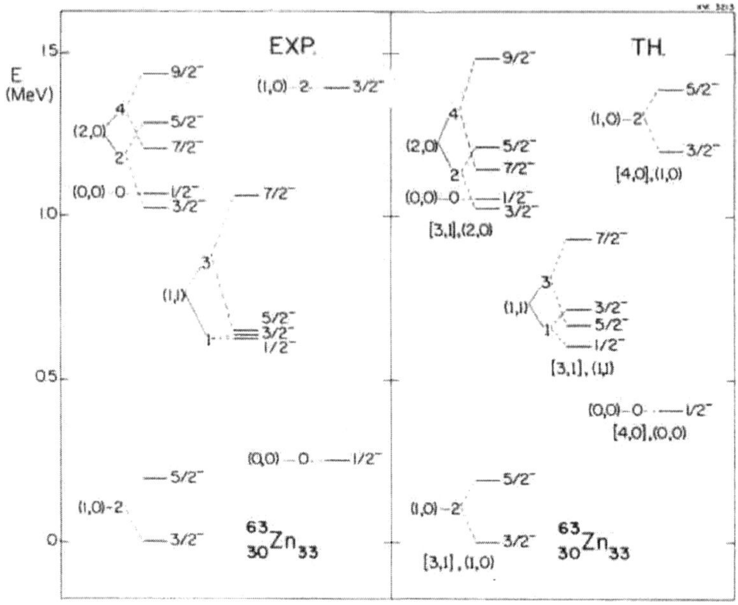

Figure 3. Comparison between the energy spectrum of the negative parity levels in the nucleus ^{63}Zn and that obtained for the $U(36/24)$ supersymmetry [7].

shown [21] that the lowlying states of ^{58}Cu, ^{60}Zn and ^{60}Ni have good pseudo-$SU(4)$ symmetry. Since the boson and fermion chains have the orbital group $U_L(6)$ and the (pseudo)spin-isospin group $SU_{ST}(4)$ in common, they may be combined into the supersymmetry

$$U(36/24) \supset U^B(36) \otimes U^F(24) \supset \cdots \supset U_L^{BF}(6) \otimes SU_{ST}^{BF}(4) \supset \cdots \quad (10)$$

Fig. 3 shows the nucleus ^{63}Zn as a candidate of a supersymmetry in the pf-shell [7].

5. Summary, conclusions and outlook

In this contribution, I have discussed how symmetry methods and algebraic models can be used to interpret and help understand the spectroscopic properties of atomic nuclei. Nuclear supersymmetry was taken as a specific example. Even though most applications have been found in medium- and heavy-mass nuclei, there are some interesting possibilities for light nuclei as well, especially for the nuclei beyond ^{56}Ni in the beginning of the pf-shell where the pseudo-$SU(4)$ symmetry is expected to be valid.

Dynamical supersymmetries provide a set of closed expressions for energies, selection rules for electromagnetic transitions and transfer reactions which may be used as benchmarks to study and interpret the experimental data, even if these symmetries may be valid only in an approximate way. In such a scheme, a supermultiplet consists of pairs or quartets of nuclei, whose properties are described simultaneously by the same form of the Hamiltonian, electromagnetic transition operators and transfer operators. Therefore, nuclear SUSY predicts explicit correlations between energies and electromagnetic transition rates in different nuclei as well as between different nucleon transfer reactions which provide a challenge and motivation for future experiments.

The concept of supersymmetry is more general than that of dynamical supersymmetry discussed in this contribution. The combination with dynamical symmetries has limited the study of nuclear SUSY, since dynamical symmetries are rather scarce and only occur in certain areas of the nuclear mass table. An example of nuclear SUSY without dynamical symmetry is a study of the Ru and Rh isotopes, in which an excellent description of the data was obtained by a combination of the $U(5)$ and $SO(6)$ dynamical symmetries [22]. This opens up the possibility to generalize nuclear SUSY to transitional regions of the nuclear mass table, and to extend the search for correlations as a result of supersymmetry.

It goes without saying that symmetry approaches as described in this contribution, *ab initio* methods, mean-field techniques and nuclear effective field theory go hand in hand, and provide complementary information about the complex dynamics of the nuclear many-body problem.

Acknowledgments

The results for the correlations in the supersymmetric Pt and Au nuclei were obtained in collaboration with José Barea and Alejandro Frank. This work was supported in part by CONACyT.

References

1. See *e.g. RIA Theory Bluebook: A Road Map* (September 2005), and various contributions to these proceedings.
2. F. Iachello, *Phys. Rev. Lett.* **44**, 772 (1980).
3. F. Iachello and A. Arima, *The interacting boson model*, (Cambridge University Press, 1987).
4. F. Iachello and O. Scholten, *Phys. Rev. Lett.* **43**, 679 (1979).

5. F. Iachello and P. Van Isacker, *The interacting boson-fermion model*, (Cambridge University Press, 1991).
6. A.B. Balantekin, I. Bars and F. Iachello, *Phys. Rev. Lett.* **47**, 19 (1981); A.B. Balantekin, I. Bars and F. Iachello, *Nucl. Phys.* A **370**, 284 (1981).
7. R. Bijker, Ph.D. thesis, University of Groningen (1984).
8. J.A. Cizewski, R.F. Casten, G.J. Smith, M.L. Stelts, W.R. Kane, H.G. Börner and W.F. Davidson, *Phys. Rev. Lett.* **40**, 167 (1978); A. Arima and F. Iachello, *Phys. Rev. Lett.* **40**, 385 (1978).
9. A.B. Balantekin, I. Bars, R. Bijker and F. Iachello, *Phys. Rev.* C **27**, 1761 (1983).
10. P. Van Isacker, J. Jolie, K. Heyde and A. Frank, *Phys. Rev. Lett.* **54**, 653 (1985).
11. A. Metz, J. Jolie, G. Graw, R. Hertenberger, J. Gröger, C. Günther, N. Warr and Y. Eisermann, *Phys. Rev. Lett.* **83**, 1542 (1999).
12. J. Gröger *et al.*, *Phys. Rev.* C **62**, 064304 (2000).
13. J. Barea, R. Bijker and A. Frank, *J. Phys. A: Math. Gen.* **37**, 10251 (2004).
14. M.L. Munger and R.J. Peterson, *Nucl. Phys.* A **303**, 199 (1978).
15. G. Graw, private communication.
16. H.-F. Wirth *et al.*, *Phys. Rev.* C **70**, 014610 (2004).
17. J. Barea, R. Bijker and A. Frank, *Phys. Rev. Lett.* **94**, 152501 (2005).
18. R. Bijker and O. Scholten, *Phys. Rev.* C **32**, 591 (1985); R. Bijker and V.K.B. Kota, *Phys. Rev.* C **37**, 2149 (1988)
19. S. Szpikowski, P. Kłosowski and L. Próchniak, *Nucl. Phys.* A **487**, 301 (1988).
20. J.P. Elliott and J.A. Evans, *Phys. Lett.* B **101**, 216 (1981).
21. P. Van Isacker, O. Juillet and F. Nowacki, *Phys. Rev. Lett.* **82**, 2060 (1999).
22. A. Frank, P. Van Isacker and D.D. Warner, *Phys. Lett.* B **197**, 474 (1987).

FERMION SYSTEMS WITH FUZZY SYMMETRIES
(LEVERAGING THE KNOWN TO UNDERSTAND THE UNKNOWN)

J. P. DRAAYER[1], K. D. SVIRATCHEVA[1],
T. DYTRYCH[1], C. BAHRI[1], K. DRUMEV[1], J. P. VARY[2]
[1]*Department of Physics and Astronomy, Louisiana State University, Baton Rouge, Louisiana 70803, USA*
[2]*Department of Physics and Astronomy, Iowa State University, Ames, IA 50011, USA*

Microscopic models, which embody the simplicity and significance of a dynamical symmetry approach to nuclear structure, are reviewed. They can reveal striking features of atomic nuclei when a symmetry dominates and solutions in domains that may otherwise be unreachable.

1. Overview of algebraic fermion models. A theory that invokes group symmetries is driven by an expectation that the wave functions of the quantum mechanical system under consideration can be characterized by their invariance properties under the corresponding symmetry transformations. But even when the symmetries are not exact, if one can find near invariant operators, the associated symmetries can be used to help reduce the dimensionality of a model space to a tractable size. Throughout the years, group-theoretical approaches have identified fundamental symmetries in light to heavy nuclei and achieved a reasonable reproduction of experimental data (for a review of fermion models, see [1]). In addition, they provide theoretical predictions for nuclear systems including heavy unstable nuclei not yet explored, and 'exotic' nuclei, such as neutron-deficient or $N \approx Z$ nuclei on the path of the nucleosynthesis *rp*-processes.

It is well-known that effective two-body interactions in nuclei are dominated by pairing and quadrupole terms. The former gives rise to a pairing gap in nuclear spectra, and the latter is responsible for enhanced electric quadrupole transitions in collective rotational bands. Indeed, within the framework of the harmonic oscillator shell-model, both limits have a clear algebraic structure in the sense that the spectra exhibit a dynamical symmetry. In the pairing limit the symplectic Sp(4) ($\sim$ SO(5)[2,3]) group together with its dual Sp(2Ω), for 2Ω shell degeneracy, use the seniority quantum number[4,5] to classify the spectra. On the other hand, in the quadrupole limit the SU(3) (sub-) structure[6] governs a shape-determined dynamics.

In light deformed nuclei, $A \lesssim 28$, the Elliot's SU(3) model[6], which incorporates the particle quadrupole and angular momentum operators, proved successful for a microscopic description of collective modes. Indeed, SU(3) is the exact symmetry group of the spherical oscillator, which is a reasonable approximation for the average potential experienced by nucleons in nuclei. Also, SU(3) is the dynamical symmetry group of the deformed oscillator, when, as is usually the case, the deformation is generated by quadrupole interactions. In many cases, a single-irrep or few-irreps calculations suffice to achieve good agreement with experimental rotational energy spectra and electromagnetic transitions (e.g., see [7,8]).

Limitations due to the fact that the SU(3) model is applied within a shell and in turn requires effective charges for transition strengths are overcome in the $\mathrm{Sp}(3,\mathbb{R})$ symplectic shell model for light nuclei (for a review see [9]). It embeds the SU(3) symmetry and in addition introduces important inter-shell excitations, including high-$\hbar\omega$ correlations and core excitations. The symplectic shell model is a microscopic formulation of the Bohr-Mottelson collective geometric model with a direct relation between a second- and a third-order scalar products of the quadrupole operator and the (β, γ)-shape variables. The $\mathrm{Sp}(3,\mathbb{R})$ symplectic model provides a microscopic description of monopole and quadrupole collective modes in deformed nuclei and a reproduction of experimental rotational energy spectra and electromagnetic transitions without effective charges (e.g., see [10,11]).

Furthermore, in the domain of light nuclei one can combine the $\mathrm{Sp}(3,\mathbb{R})$ symplectic shell model and the no-core shell model (NCSM)[12] to push forward the present frontiers in nuclear structure physics. The NCSM+$\mathrm{Sp}(3,\mathbb{R})$ allows us to use modern realistic interactions without any approximation (for the interaction and the size of the model space) for low-$\hbar\omega$ configurations and hence to fully account for important short- and intermediate-range correlations, while selecting only dominant high-$\hbar\omega$ basis states responsible for multi-shell development of collective motion.

In the region of medium mass nuclei around the $N = Z$ line (currently explored by radioactive beam experiments) protons and neutrons occupy the same major shell and hence their mutual interactions are expected to strongly influence the structure and decay modes of such nuclei. In addition to like-particle (*pp* and *nn*) pairing correlations the close interplay of *pp*, *nn* and proton-neutron (*pn*) pairs and the isospin-symmetry influence are microscopically described by the Sp(4) pairing model[13].

For heavy nuclei ($A \gtrsim 100$), the discovery of the pseudo-spin symmetry[14,15] and its fundamental nature[16,17,18] establishes the pseudo-SU(3) model[19]. The pseudo-spin scheme is an excellent starting point for a many-particle description of heavy nuclei, whether or not they are deformed. As for the SU(3) shell model, in many cases leading-irrep calculations (e.g., see [20]) or mixed-irrep calculations (e.g., see [21]) achieve good agreement with experimental data. The pseudo-SU(3) shell model provides a further understanding of the $M1$ transitions in nuclei such as the even-even $^{160-164}$Dy and $^{156-160}$Gd isotopes, specifically it reflects on the scissors and twist modes as well as the observed fragmentation, that is, the break-up of the $M1$ strength among several levels closely clustered around a few strong transition peaks in the 2-4 MeV energy region[22].

In medium-mass and heavy nuclei, where the pseudo-spin scheme can be applied to the normal parity orbitals and valence spaces are intruded by a unique parity highest-j orbit from the shell above, a major step towards understanding the significance of the intruder level is achieved by a pseudo-SU(3) plus intruder level shell model[23] that is currently under development[24].

Furthermore, the advantages of the symplectic Sp(3, $\mathbb{R}$) extension of the SU(3) model can be employed beyond the light nuclei domain towards a description of heavy nuclei in the framework of the pseudo-Sp(3, $\mathbb{R}$) shell model (e.g., see [25]). While early developments demonstrate the potential of such a model for studying the structure of heavy nuclear systems, it has not been fully explored.

In what follows, we present some recent results for three algebraic fermion models where the symmetries are fuzzy – meaning not exact – but nevertheless extremely useful in gaining a deeper understanding of the structure of real nuclei.

2. Symplectic Sp(4) **pairing model.** An algebraic approach, with Sp(4) the underpinning symmetry and with only six parameters, can be used to provide a reasonable microscopic description of pairing-governed 0^+ states in a total of 306 even-even and odd-odd nuclei with mass $40 \leq A \leq 100$ where protons and neutrons are filling the same major shell[26,13]. We employ the most general Hamiltonian with Sp(4) dynamical symmetry,

$$H_{\mathrm{sp}(4)} = -G\sum_{i=-1}^{1} \hat{A}_i^\dagger \hat{A}_i - F\hat{A}_0^\dagger \hat{A}_0 - \frac{E}{2\Omega}(\hat{T}^2 - \frac{3\hat{N}}{4}) \\ - D(\hat{T}_0^2 - \frac{\hat{N}}{4}) - C\frac{\hat{N}(\hat{N}-1)}{2} - \epsilon\hat{N}, \qquad (1)$$

where G, F, E, D, C and $\epsilon > 0$ are parameters (refer to Table I in [13] for their estimates). In (1), $\hat{N}$ counts the total number of particles, $\hat{T}^2$ is the isospin operator and the $\hat{A}^{\dagger}_{0,+1,-1}$ group generators, which build the basis states by a consequent action on the vacuum state (a core like ^{40}Ca or ^{56}Ni), create, respectively, a proton-neutron (pn) pair, a proton-proton (pp) pair or a neutron-neutron (nn) pair of total angular momentum $J^{\pi} = 0^+$ and isospin $T = 1$. The model Hamiltonian (1) represents an effective microscopic interaction that conserves the T_0 third projection of the isospin and includes proton-neutron and like-particle isovector ($T = 1$) pairing plus symmetry terms (the latter is related to a proton-neutron isoscalar ($T = 0$) force). The model interaction (1) is found to correlate strongly with realistic interactions like CD-Bonn+3 terms[27] in the $1f_{7/2}$ region[28] and reflects a large portion of the GXPF1 realistic interaction[29] in the upper-fp shell. In addition, the D-term in (1) introduces isospin symmetry breaking and the F-term accounts for a plausible, but weak, isospin mixing[30]. Both terms are significant in non-analog β-decays studies and also yield quantitative results that are better (e.g., by 85% in $1f_{7/2}$) than the ones with $F = D = 0$.

Good agreement with experiment (small χ-statistics) is observed (Fig. 1) when theoretical eigenvalues of (1) are compared to Coulomb-corrected[31] experimental energies[32]. The theory predicts the lowest isobaric analog 0^+ state energy with a deviation ($\chi/\Delta E_{0,exp} \times 100[\%]$) of 0.5% for $1f_{7/2}$ and $1f_{5/2}2p_{1/2}2p_{3/2}1g_{9/2}$ nuclei in the corresponding energy range considered, $\Delta E_{0,exp}$. The model estimates the binding energy of the proton-rich ^{48}Ni nucleus to be 348.1 MeV, which is 0.07% greater than the sophisticated semi-empirical estimate[33] and only 4% away from the experimental value reported later-on[34]. The ^{68}Se waiting-point nuclide along the rp-path is estimated to be 574.3MeV, only 2.7% away from the 2004 precise mass measurement[35]. Likewise, for the odd-odd nuclei with energy spectra not yet measured the theory predicts the energy of their lowest isobaric analog 0^+ state: 358.62 MeV (^{44}V), 359.34 MeV (^{46}Mn), 357.49 MeV (^{48}Co), 394.20 MeV (^{50}Co) (Fig. 1, right). The Sp(4) model predicts the relevant 0^+ state energies for an additional 165 even-A nuclei in the medium mass region (Fig. 1, left). The binding energies for 25 of them are also calculated in [33]. For these even-even nuclei, we predict binding energies that on average are 0.05% less than the semi-empirical approximation[33].

Furthermore, without any parameter variation, the theoretical energy spectra of the isobaric analog 0^+ states are found to agree remarkably well with the experimental values where data is available[13]. This agreement represents a valuable result because the higher-lying 0^+ states under

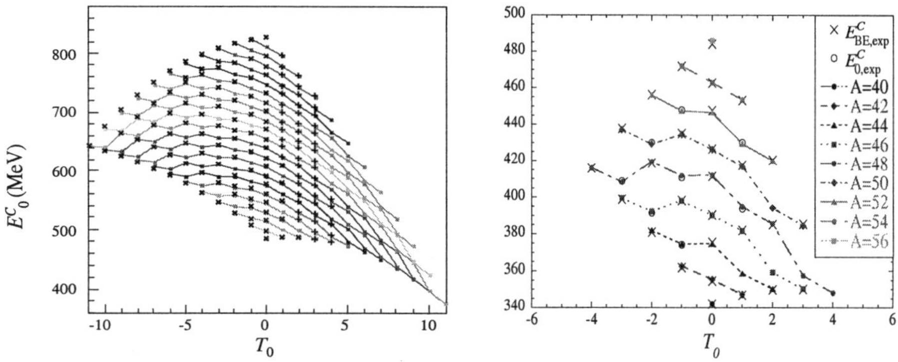

Figure 1. Theoretical isobaric analog 0^+ state energies in MeV of isobaric sequences (lines) (including the Coulomb energy) vs. the isospin projection T_0. Left: isobars with $A = 56, 58, \dots, 100$ in the $1f_{5/2}2p_{1/2}2p_{3/2}1g_{9/2}$ shell (^{56}Ni core), compared to experimental values (black '×') and semi-empirical estimates by P. Möller *et al.* (blue '+'). Right: isobars in the $1f_{7/2}$ level compared to experimental binding energies (×) or energies of the lowest isobaric analog 0^+ excited states (○).

consideration constitute an experimental set independent of the data that enters the statistics to determine the model parameters in (1).

In addition, we examine the detailed features of nuclei by discrete derivatives of the energy function (1) filtering out the strong mean-field influence[36]. This investigation reveals a remarkable reproduction of the two-proton S_{2p} and two-neutron S_{2n} separation energies, the irregularities found around the $N = Z$ region, the like-particle and pn isovector pairing gaps, and a prominent staggering behavior observed between groups of even-even and odd-odd nuclides[36]. The zero point of S_{2p} along an isotone sequence determines the two-proton-drip line, which according to the $Sp(4)$ model for the $1f_{5/2}2p_{1/2}2p_{3/2}1g_{9/2}$ shell lies near the following even-even nuclei[36]: $^{60}Ge_{28}$, $^{64}Se_{30}$, $^{68}Kr_{32}$, $^{72}Sr_{34}$, $^{76}Zr_{36}$, $^{78}Zr_{38}$, $^{82}Mo_{40}$, $^{86}Ru_{42}$, $^{90}Pd_{44}$, $^{94}Cd_{46}$, beyond which the higher-Z isotones are unstable with respect to diproton emissions in close agreement with other estimates[37,33,38] despite the lack of experimental data. In addition, we find a small quadratic mean of the difference in S_{2p} between our model and the other theoretical predictions where data is available, namely, 0.32 MeV in comparison with [37], 0.78 MeV with [33] and 0.43 MeV with [38].

While the model describes only isobaric analog 0^+ states of even-A medium mass nuclei with protons and neutrons in the same shell, it reveals a fundamental feature of the nuclear interaction, which governs these states. Namely, the latter possesses a clear Sp(4) dynamical symmetry.

3. Pseudo-SU(3) plus intruder level model. The role of intruder levels that penetrate down into lower-lying shells in atomic nuclei has been the focus of many studies and debates. These levels are found in heavy deformed nuclei where the strong spin-orbit interaction destroys an underlying harmonic oscillator symmetry of the nuclear mean-field potential.

We carry out m-scheme shell-model calculations for the ^{58}Cu and ^{64}Ge nuclei in the $1f_{5/2}2p_{1/2}2p_{3/2}1g_{9/2}$ model space assuming the occupancy of the $f_{7/2}$ orbital to be 'frozen'. This choice was motivated by the $f_{7/2}$ orbit's high occupation as reported elsewhere[39]. The Hamiltonian we used is a G-matrix with a phenomenologically adjusted monopole part[40]. A renormalized version of this interaction in the $pf_{5/2}$ space has been introduced for describing beta decays[41].

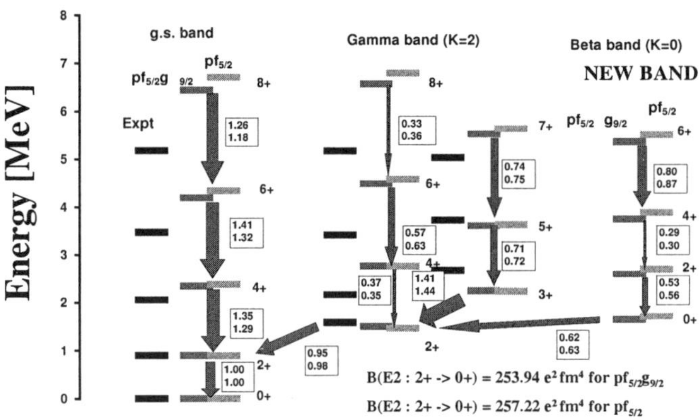

Figure 2. Energy spectrum and B(E2) transition strengths for ^{64}Ge. The width of the arrows in the figure represent the relative B(E2) strengths, normalized to unity for the ground band $2^+ \rightarrow 0^+$ transition. Numbers in each box are for the $1f_{5/2}2p_{1/2}2p_{3/2}1g_{9/2}$ and the restricted upper *fp* model space, respectively.

Results for the energy spectrum and B(E2) strengths of ^{64}Ge are shown in Fig. 2 for both model spaces. The renormalized version of the theory in the upper *fp* subspace not only reproduces the excitation energies obtained in the larger $1f_{5/2}2p_{1/2}2p_{3/2}1g_{9/2}$ space, but also gives very similar values for the B(E2) transition strengths. These results also confirm those from a study using a schematic interaction[42]. Similar behavior was observed for ^{58}Cu. Besides the ground state and gamma bands for ^{64}Ge a new (possibly beta) band is identified.

In addition, after rescaling occupancies of the states obtained in the upper *fp* subspace, a pattern that is very similar to that of the occupancies in the larger $1f_{5/2}2p_{1/2}2p_{3/2}1g_{9/2}$ space is obtained.

In short, novel shell-model calculations for ^{58}Cu and ^{64}Ge in the $1f_{5/2}2p_{1/2}2p_{3/2}1g_{9/2}$ model space using a realistic interaction are compared to those generated by an appropriately renormalized counterpart of the interaction in the truncated upper *fp* subspace. The results suggest that reliable computations can be performed in a space that does not explicitly include the intruder level as long as the interaction and the transition operators are renormalized appropriately.

4. No-core shell model plus Sp(3, ℝ) **extension.** The symplectic shell model is based on the noncompact symplectic $\mathfrak{sp}(3,\mathbb{R})$ algebra with a subalgebraic structure that gives rise to rich underlying physics for a microscopic description of multiple collective modes in nuclei. This follows from the fact that the mass quadrupole and monopole moments operators, the many-particle kinetic energy, the angular and vibrational momenta are all elements of the $\mathfrak{sp}(3,\mathbb{R}) \subset \mathfrak{su}(3)$ algebraic structure[9].

The symplectic basis states, $|\Gamma_\sigma \Gamma_n \rho \Gamma_\omega \kappa (LS) J M_J\rangle$, are constructed by acting with polynomials in the symplectic raising operator, $A^{(2\,0)}$, on a set of basis states of a symplectic bandhead, $|\Gamma_\sigma\rangle$. They are labeled according to the reduction chain

$$\begin{array}{ccccc} \mathrm{Sp}(3,\mathbb{R}) & \supset & \mathrm{U}(3) & \supset & \mathrm{SO}(3) \\ \Gamma_\sigma & \Gamma_n \rho & \Gamma_\omega & \kappa & L \end{array}$$

where $\Gamma_\sigma \equiv N_\sigma\,(\lambda_\sigma, \mu_\sigma)$ labels Sp(3, ℝ) irreducible representations. The $\Gamma_n \equiv n\,(\lambda_n, \mu_n)$ set of quantum numbers gives the overall SU(3) coupling of $n/2$ raising operators acting on $|\Gamma_\sigma\rangle$. $\Gamma_\omega \equiv N_\omega\,(\lambda_\omega, \mu_\omega)$ specifies the SU(3) symmetry of a symplectic state and $N_\omega = N_\sigma + n$ is the total number of oscillator quanta related to the eigenvalue, $N_\omega \hbar\Omega$, of a center-of-mass motion free harmonic-oscillator (HO) Hamiltonian. The basis states of a Sp(3, ℝ) irreducible representation can be expanded in HO (m-scheme) basis, which is the basis utilized by the no-core shell-model (NCSM)[12].

In the case of ^{12}C one can construct 13 unique so-called $0\hbar\Omega$-Sp(3, ℝ) irreducible representations. $0\hbar\Omega$ means that the symplectic bandhead basis states, $|\Gamma_\sigma\rangle$, lie within $0\hbar\Omega$ many-particle harmonic-oscillator space. For each of the $0\hbar\Omega$-Sp(3, ℝ) irreducible representations we generate basis states up to $N_{max} = 6$ ($6\hbar\Omega$), which is the current limit for NCSM calculations. Typical the dimension of a symplectic representation is of order 10^3, comparing to 10^7 in the case of NCSM m-scheme basis space.

Table 1. Probability distribution of the three most important $\mathrm{Sp}(3,\mathbb{R})$ irreps and of the NCSM wavefunctions.

	$0\hbar\Omega$	$2\hbar\Omega$	$4\hbar\Omega$	$6\hbar\Omega$	Total
		J=0			
$(0\ 4)S=0$	46.26	11.39	4.15	1.11	62.91
$(1\ 2)S=1$	4.80	1.87	0.80	0.30	7.77
$(1\ 2)S=1$	4.72	1.84	0.79	0.29	7.64
$\mathrm{Sp}(3,\mathbb{R})$ Total	55.78	15.10	5.74	1.70	78.32
NCSM	56.18	22.40	12.81	7.00	98.38
		J=2			
$(0\ 4)S=0$	46.80	11.33	3.99	1.06	63.18
$(1\ 2)S=1$	4.84	1.65	0.69	0.25	7.43
$(1\ 2)S=1$	4.69	1.60	0.67	0.24	7.20
$\mathrm{Sp}(3,\mathbb{R})$ Total	56.33	14.58	5.35	1.55	77.81
NCSM	56.63	21.79	12.73	7.28	98.43
		J=4			
$(0\ 4)S=0$	51.45	11.23	3.71	0.94	67.33
$(1\ 2)S=1$	3.04	0.89	0.35	0.12	4.40
$(1\ 2)S=1$	3.01	0.88	0.35	0.12	4.36
$\mathrm{Sp}(3,\mathbb{R})$ Total	57.50	13.00	4.41	1.18	76.09
NCSM	57.64	20.34	12.59	7.66	98.23

The lowest-lying eigenstates of ^{12}C are calculated by NCSM approach using the Many Fermion Dynamics (MFD) code[43] with the effective interaction derived from the realistic JISP16 *NN* potential for oscillator strengths of $\hbar\Omega = 15$ MeV. The large overlaps of the symplectic states with the NCSM wave functions for 0, 2, 4 and $6\hbar\Omega$ subspaces of the m-scheme basis (Tables 1) reveal that around 80% of the latter are symmetric under $\mathrm{Sp}(3,\mathbb{R})$ transformations. Apparently, for all three wave functions, the highest contribution comes from the leading, most deformed, $(0\,4)S = 0$ $\mathrm{Sp}(3,\mathbb{R})$ irreducible representation. This contribution gets higher towards $J = 4_1^+$, where mixing due to other, less deformed, configurations decreases.

Clearly, the 0^+, 2^+ and 4^+ states, which are constructed in terms of the three $\mathrm{Sp}(3,\mathbb{R})$ irreps with probability amplitudes defined by the overlaps with the NCSM wavefunctions, can be used as a quite good approximation for a microscopic description of the 0^+_{gs}, 2_1^+ and 4_1^+ states in ^{12}C. Within this assumption, the $B(E2: 2_1^+ \to 0^+_{g.st.})$ value turns out to be as much as 81% of the NCSM estimate.

In short, the low-lying states in ^{12}C are quite well explained by only three $\mathrm{Sp}(3,\mathbb{R})$ irreps of 1098 symplectic states, that is only 0.003% of the

NCSM space dimension, with a dominance of the most deformed $(0\ 4)S = 0$ collective configuration. Our findings, as a 'proof-of-principle', suggests that a NCSM+Sp$(3,\mathbb{R})$ structure could allow one to extend no-core calculations to higher $\hbar\omega$ and heavier nuclei.

In summary, models based on exact or just good (broken but dominant) symmetries in fermion systems can play a significant role in our understanding low-lying nuclear structure; specifically, as shown here, in the development of collective rotational motion and the formation of correlated pairs in nuclei. They also allow us to truncate a model space to typically only a fraction the size encountered in models that do not exploit what we have dubbed here as fuzzy symmetries.

Acknowledgments

This work was supported by the US National Science Foundation, Grant Numbers 0140300 & 0500291 and the Southeastern Universities Research Association (SURA).

References

1. J. P. Draayer, *Fermion Models*, in *Algebraic Approaches to Nuclear Structure*, ed. R. Casten, (Harwood Academic Publishers), Ch. 7, 423 (1992).
2. K. Helmers, *Nucl. Phys.* **23**, 594 (1961).
3. K. T. Hecht, *Nucl. Phys.* **63**, 177(1965); *Phys. Rev.* **139**, B794 (1965); *Nucl. Phys.* **A102**, 11 (1967); J. N. Ginocchio, *Nucl. Phys.* **74**, 321 (1965).
4. G. Racah, *Phys. Rev.* **62**, 438 (1942); *Phys. Rev.* **63**, 367 (1943).
5. B. H. Flowers, *Proc. Roy. Soc.* (London) **A212**, 248 (1952).
6. J. P. Elliott, *Proc. Roy. Soc.* (London) **A245**, 128 (1958); **A245**, 562 (1958); J. P. Elliott and M. Harvey, *Proc. Roy. Soc.* (London) **A272**, 557 (1963).
7. H.A. Naqvi and J.P. Draayer, *Nucl. Phys.* **A516**, 351 (1990).
8. C.E. Vargas, J.G.Hirsch, and J.P.Draayer, *Nucl. Phys.* **A690**, 409 (2001) .
9. D.J. Rowe, *Rep. Prog. Phys.* **48**, 1419 (1985).
10. G. Rosensteel and D.J. Rowe, *Phys. Rev. Lett.* **38**, 10 (1977); *Ann. Phys. N Y* **126**, 343 (1980).
11. J.P. Draayer, K.J. Weeks, and G. Rosensteel, *Nucl. Phys.* **A 413**, 215 (1984).
12. P. Navrátil, J.P. Vary, and B.R. Barrett, *Phys. Rev. Lett.* **84**, 5728 (2000).
13. K. D. Sviratcheva, A. I. Georgieva, and J. P. Draayer, *Phys. Rev.* **C70**, 064302 (2004).
14. K. T. Hecht and A. Adler, *Nucl. Phys.* **A137**, 129 (1969).
15. A. Arima A, M. Harvey, and M K. Shimizu, *Phys. Lett.* **30B**, 517 (1969).
16. J.P. Draayer, in *Proceedings of Symmetries in Physics*, Cocoyoc, Mexico, June 3-7, 1991, ed. A. Frank and K.B. Wolf (Springer-Verlag, Berlin).
17. C. Bahri, J. P. Draayer, and S.A. Moszkowski, *Phys. Rev. Lett.* **68**, 2133 (1992).

18. A.L. Blokhin, C. Bahri, and J.P. Draayer, *Phys. Rev. Lett.* **74**, 4149 (1995).
19. R.D. Ratna Raju, J.R Draayer, and K.T. Hecht, *Nucl. Phys.* **A202**, 433 (1973).
20. J.R Draayer and K.J. Weeks, *Phys. Rev. Lett.* **51**, 1422 (1983).
21. G. Popa, J. G. Hirsch, and J. P. Draayer, *Phys. Rev.* **C62**, 064313 (2000).
22. T. Beuschel, J. P. Draayer, D. Rompf, and J. G. Hirsch, *Phys. Rev.* **C57**, 1233 (1998).
23. K.J. Weeks, C.S. Han, and J.P. Draayer, *Nucl. Phys.* **A371**, 19 (1981).
24. K. P. Drumev, C. Bahri, V. G. Gueorguiev and J. P. Draayer, in *Proceedings of Nuclear Physics, Large and Small: International Conference on Microscopic Studies of Collective Phenomena*, Morelos, Mexico, April 19-22, 2004, eds. R. Bijker, R.F. Casten, and A. Frank, Melville, New York, *AIP Conference Proceeding* **726**, 219 (2004).
25. O. Castaños, P.O. Hess, J.P. Draayer, and P. Rochford, *Nucl. Phys.* **A524**, 469 (1991).
26. K. D. Sviratcheva, A. I. Georgieva, and J. P. Draayer, *J. Phys. G: Nucl. Part. Phys.* **29**, 1281 (2003).
27. S. Popescu, S. Stoica, J. P. Vary, and P. Navratil, to be published.
28. K. D. Sviratcheva, J. P. Draayer, and J. P. Vary, *Phys. Rev.* **C73**, 034324 (2006).
29. M. Honma, T. Otsuka, B. A. Brown, and T. Mizusaki, *Phys. Rev.* **C69**, 034335 (2004).
30. K. D. Sviratcheva, A. I. Georgieva, and J. P. Draayer, *Phys. Rev.* **C72**, 054302 (2005).
31. J. Retamosa, E. Caurier, F. Nowacki and A. Poves, *Phys. Rev.* **C55**, 1266 (1997).
32. G. Audi and A. H. Wapstra, *Nucl. Phys.* **A595**, 409 (1995); R. B. Firestone and C. M. Baglin, *Table of Isotopes*, 8th Edition (John Wiley & Sons, 1998).
33. P. Möller, J. R. Nix and K.-L. Kratz, *LA-UR-94-3898* (1994); *At. Data Nucl. Data Tables* **66**, 131 (1997).
34. B. Blank *et al.*, *Phys. Rev. Lett.* **84**, 1116 (2000).
35. J.A. Clark *et al.*, *Phys. Rev. Lett.* **92**, 192501 (2004).
36. K. D. Sviratcheva, A. I. Georgieva, and J. P. Draayer, *Phys. Rev.* **C69**, 024313 (2004).
37. E. Ormand, *Phys. Rev.* **C55**, 2407 (1997).
38. B. A. Brown, R. R. C. Clement, H. Schatz, and A. Volya, *Phys. Rev.* **C65**, 045802 (2002).
39. M. Honma, T. Mizusaki, and T. Otsuka, *Phys. Rev. Lett.* **77**, 3315 (1996).
40. E. Caurier, F. Novacki, A. Poves, and J. Retamosa, *Phys. Rev. Lett.* **77**, 1954 (1996).
41. P. Van Isacker, O. Juillet, and F. Nowacki, *Phys. Rev. Lett.* **82**, 2060 (1999).
42. K. Kaneko, M. Hasegawa, and T. Mizusaki, *Phys. Rev.* **C66**, 051306(R) (2002).
43. J. P. Vary and D. C. Zheng, *"The Many-Fermion-Dynamics Shell-Model Code"*, Iowa State University, 1994 (unpublished).

PROTON-NEUTRON ASYMMETRY IN EXOTIC NUCLEI

M. A. CAPRIO
Center for Theoretical Physics, Yale University,
New Haven, Connecticut 06520-8124, USA
E-mail: mark.caprio@yale.edu

The possibility of proton-neutron asymmetric collective structure in exotic nuclei is considered. The signatures of and conditions for proton-neutron triaxial deformation are discussed within the framework of the proton-neutron interacting boson model.

1. Introduction

An exotic beam facility such as the Rare Isotope Accelerator will yield an extensive new set of nuclei. These nuclei will differ from familiar nuclei near stability in having a substantial proton-neutron imbalance, and they might also consequently differ in underlying shell structure.[1] As part of the theoretical effort in preparation for such a facility, it is necessary to anticipate new collective phenomena which might arise from these differences and to deduce signatures by which these phenomena can be recognized. It is also necessary to determine in which nuclei such phenomena may be expected to occur.

Most low-energy collective phenomena in nuclei near stability are essentially *isoscalar*, in that the proton and neutron distributions in the ground state are similar. An underlying reason for this is that deformation arises from a strong proton-neutron quadrupole interaction, which tends to couple the proton and neutron deformations. Proton-neutron asymmetry is instead manifest in a more subtle form, in collective excitations modes. These include the scissors mode,[2–4] mixed-symmetry oscillations,[5,6] giant and pygmy dipole resonances,[7,8] and, in odd-odd nuclei, asymmetric coupling of the unpaired nucleons to the collective core[9] (Fig. 1).

The question remains as to whether or not proton-neutron asymmetry can be present in the ground state. In very neutron-rich nuclei there will be a gross imbalance in proton and neutron numbers. Moreover, the proton and neutron valence spaces will be well separated. This could lead to reduced proton-neutron coupling strengths and therefore a larger role for proton-neutron asymmetry in the

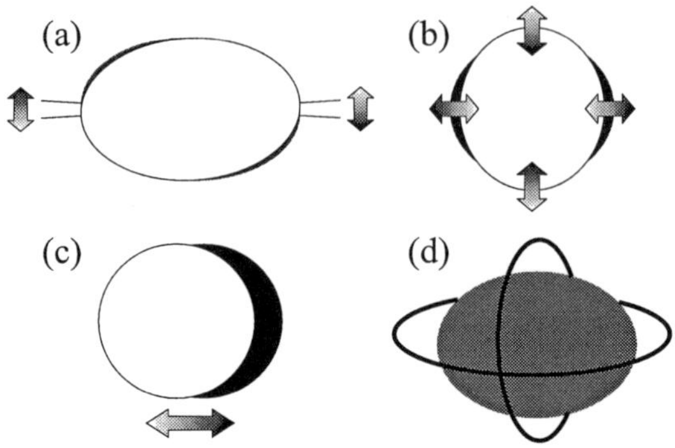

Figure 1. Schematic illustrations of proton-neutron asymmetric collective excitation modes: (a) scissors mode, (b) mixed-symmetry mode, (c) electric dipole resonance, and (d) asymmetric coupling in odd-odd systems.

ground state. Such asymmetry would have major consequences for the ground state properties, excitation modes, and transition radiations (in particular, asymmetric proton and neutron motion leads to electric or magnetic dipole transitions).

Proton-neutron asymmetry is intimately related to triaxial deformation. Triaxial deformation can arise in an ordinary proton-neutron *symmetric* system, due either to higher-order interactions in the Hamiltonian[10] or to the presence of high-multipolarity (hexadecapole) nucleon pairs.[11] However, triaxial structure also naturally arises in any proton-neutron *asymmetric* situation in which the proton and neutron symmetry axes are unaligned, since then there is no common symmetry axis for the nucleus as a whole [Fig. 2 (left)].

2. Dynamical symmetries

For proton-neutron symmetric nuclei, the quadrupole deformation degrees of freedom can be described using the interacting boson model[12] (IBM), in which the nucleus is treated as consisting of s-wave and d-wave nucleon pairs. However, for proton-neutron asymmetric structure, we must instead consider the proton-neutron interacting boson model[12, 13] (IBM-2), in which proton and neutron pairs are treated as distinct constituents. An IBM-2 Hamiltonian

$$H = \varepsilon_\pi \hat{n}_{d\pi} + \varepsilon_\nu \hat{n}_{d\nu} + \kappa_{\pi\pi} \hat{Q}_\pi^{\chi_\pi} \cdot \hat{Q}_\pi^{\chi_\pi} + \kappa_{\pi\nu} \hat{Q}_\pi^{\chi_\pi} \cdot \hat{Q}_\nu^{\chi_\nu} + \kappa_{\nu\nu} \hat{Q}_\nu^{\chi_\nu} \cdot \hat{Q}_\nu^{\chi_\nu} + \cdots \quad (1)$$

in general contains some terms which involve just a single constituent species ($\hat{n}_{d\pi}$, $\hat{Q}_\pi^{\chi_\pi} \cdot \hat{Q}_\pi^{\chi_\pi}$, ...) and others which yield interactions between the proton and neutron fluids ($\hat{Q}_\pi^{\chi_\pi} \cdot \hat{Q}_\nu^{\chi_\nu}$, ...) (see Ref. 12 for detailed definitions of the operators).

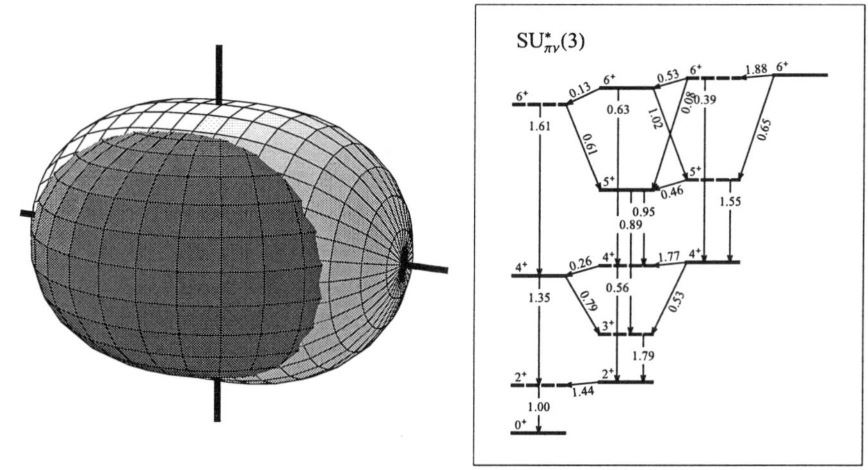

Figure 2. Proton-neutron asymmetric triaxial structure in the $\mathrm{SU}^*_{\pi\nu}(3)$ dynamical symmetry. (Left) Geometrical interpretation of the equilibrium structure. A prolate deformed proton fluid and oblate deformed neutron fluid are coupled with orthogonal symmetry axes. (Right) Low-lying excitation spectrum, showing $E2$ transition strengths within the ground state representation. Figure adapted from Refs. 20 and 21.

The IBM-2 supports several proton-neutron symmetric (isoscalar) dynamical symmetries directly analogous to those of the IBM, namely $\mathrm{U}_{\pi\nu}(5)$ (spherical oscillator), $\mathrm{SO}_{\pi\nu}(6)$ (deformed γ-soft), $\mathrm{SU}_{\pi\nu}(3)$ (prolate rotor), and $\overline{\mathrm{SU}_{\pi\nu}(3)}$ (oblate rotor).[14] However, the IBM-2 also supports qualitatively new proton-neutron asymmetric (isovector) symmetries, corresponding to the subalgebra chains

$$\mathrm{U}_\pi(6)\otimes\mathrm{U}_\nu(6)\supset\left\{\begin{array}{c}\mathrm{SU}_\pi(3)\otimes\overline{\mathrm{SU}_\nu(3)}\supset\mathrm{SU}^*_{\pi\nu}(3)\\ \overline{\mathrm{SU}_\pi(3)}\otimes\mathrm{SU}_\nu(3)\supset\overline{\mathrm{SU}^*_{\pi\nu}(3)}\end{array}\right\}\supset\mathrm{SO}_{\pi\nu}(3). \qquad (2)$$

In the $\mathrm{SU}^*_{\pi\nu}(3)$ symmetry,[15–19] a proton fluid with axially symmetric prolate deformation is coupled to a neutron fluid with axially symmetric oblate deformation, with their symmetry axes orthogonal to each other [Fig. 2 (left)]. In the $\overline{\mathrm{SU}^*_{\pi\nu}(3)}$ symmetry, the proton and neutron deformations are interchanged. In either case, the resulting overall composite nuclear shape is highly proton-neutron asymmetric and triaxially deformed.

The $\mathrm{SU}^*_{\pi\nu}(3)$ dynamical symmetry presents a variety of exotic spectroscopic features. These serve as possible signatures of proton-neutron triaxiality. Energy eigenvalues follow the usual $\mathrm{SU}(3)$ formula, and level energies within an $\mathrm{SU}^*_{\pi\nu}(3)$ representation follow $L(L+1)$ rotational spacings. However, different $\mathrm{SU}(3)$ representations are involved from those of the usual axially symmetric

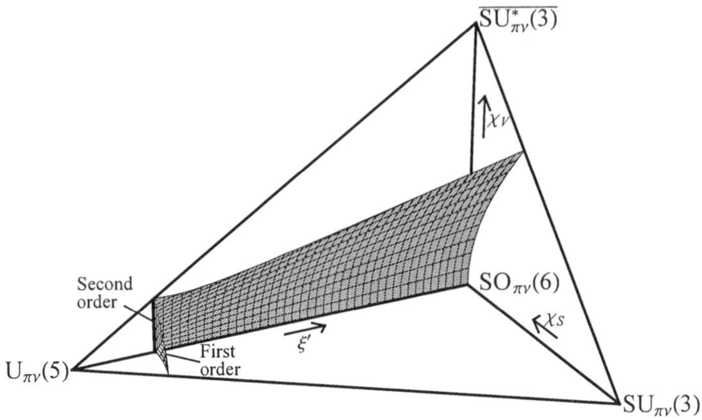

Figure 3. Phase diagram of the proton-neutron interacting boson model (IBM-2). Surfaces of first-order and second-order transition between regions of undeformed, axially symmetric deformed, and triaxially deformed equilibria are shown. Only the quadrant $\chi_S < 0$ and $\chi_V > 0$ is shown for simplicity. Figure from Ref. 25.

rotor. The $\mathrm{SU}^*_{\pi\nu}(3)$ ground state representation contains multiple degenerate rotational bands, with K quantum numbers 0, 2, 4, ... [Fig. 2 (right)]. At higher energies, excitations corresponding to coupled β and γ vibrations are present.[22] There is also an "orthogonal" scissors mode, in which the proton and neutron symmetry axes oscillate about their equilibrium perpendicular relative orientation.[17, 22]

Within the ground state representation, many of the *interband* $E2$ transitions are comparable in strength to *in-band* transitions [Fig. 2 (right)]. However, a discrete parity-like symmetry[23] imposes selection rules on the $E2$ transitions, and so some of the *in-band* transitions are in fact forbidden. The $B(E2)$ strength pattern is remarkably similar to that of the classic rigid triaxial rotor of the Davydov model.[24] The $\mathrm{SU}^*_{\pi\nu}(3)$ predictions are also characterized by the presence of strong $M1$ admixtures to the transitions within a representation, comparable in strength to the $M1$ transitions involving scissors or mixed-symmetry excitations. The $M1$ strengths arise from the large static separation of the proton and neutron fluids in the equilibrium configuration and hence large asymmetry in their motion.

3. Phase transitions

To determine the nature of the transition between proton-neutron symmetric and asymmetric structure, we consider the phase diagram of the IBM-2.[20, 25, 26] The classical coordinates ("order parameters") of the IBM-2, obtained through the coherent state formalism, are the Bohr deformation variables (β_π, γ_π, β_ν, and γ_ν) of the proton and neutron fluids and the Euler angles ($\vartheta_1, \vartheta_2, \vartheta_3$) describing the

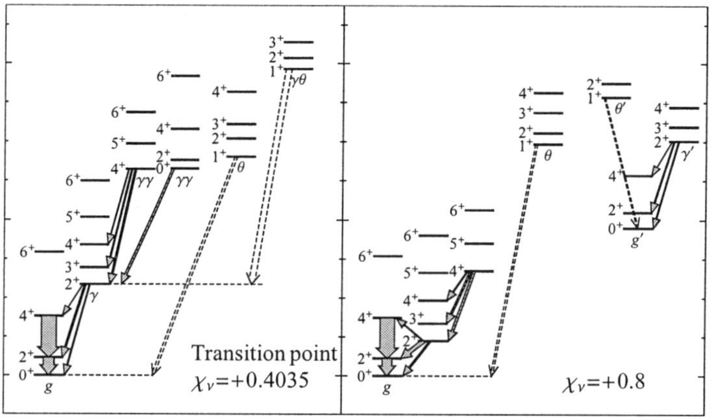

Figure 4. Level schemes for the $\mathrm{SU}_{\pi\nu}(3)$–$\mathrm{SU}^*_{\pi\nu}(3)$ transition with no Majorana operator: at the critical point (left) and beyond the critical point (right). The arrows indicate selected $E2$ transition strengths (solid) and $M1$ transition strengths (dashed). Figure adapted from Ref. 20.

relative orientations of these fluids.[27] A simple Hamiltonian which can reproduce all the dynamical symmetries of the IBM-2 is

$$H = \frac{1-\xi'}{N}(\hat{n}_{d\pi} + \hat{n}_{d\nu}) - \frac{\xi'}{N^2}(\hat{Q}_\pi^{\chi_\pi} + \hat{Q}_\nu^{\chi_\nu}) \cdot (\hat{Q}_\pi^{\chi_\pi} + \hat{Q}_\nu^{\chi_\nu}). \tag{3}$$

The parameter ξ' controls the overall tendency towards spherical or deformed nuclear shape, χ_π the tendency of the proton fluid towards prolate or oblate deformation, and χ_ν the tendency of the neutron fluid towards prolate or oblate deformation. It is convenient instead to define "scalar" and "vector" quadrupole parameters $\chi_S \equiv \frac{1}{2}(\chi_\pi + \chi_\nu)$, determining the overall prolate-oblate tendency, and $\chi_V \equiv \frac{1}{2}(\chi_\pi - \chi_\nu)$, determining the tendency towards proton-neutron asymmetry. The Hamiltonian thus has three "control parameters" (ξ', χ_S, and χ_V).

The phase structure of the IBM-2 is obtained by considering the evolution of the global minimum of the coherent state energy surface $\mathscr{E}(\beta_\pi, \gamma_\pi, \beta_\nu, \gamma_\nu, \vartheta_1, \vartheta_2, \vartheta_3)$, as discussed in further detail in Ref. 20. The problem of determining the phase diagram for the IBM-2 is more complicated than for the one-fluid IBM, and a combination of analytic and numerical methods must be used.

The resulting phase diagram is shown in Fig. 3. Surrounding $\mathrm{SU}_{\pi\nu}(3)$ is a region of parameter space in which the equilibrium deformation is axially symmetric ($\gamma_\pi = \gamma_\nu = 0°$). On a surface of second order phase transition, axially symmetric deformation gives way to proton-neutron triaxial deformation, as γ_π and/or γ_ν drifts away from $0°$. [The $\mathrm{SU}^*_{\pi\nu}(3)$ limit has $\gamma_\pi = 0°$ and $\gamma_\nu = 60°$, and $\overline{\mathrm{SU}^*_{\pi\nu}(3)}$ has $\gamma_\pi = 60°$ and $\gamma_\nu = 0°$.] In the parameter plane $\chi_V = 0$ [containing the $\mathrm{U}_{\pi\nu}(5)$, $\mathrm{SO}_{\pi\nu}(6)$, and $\mathrm{SU}_{\pi\nu}(3)$ dynamical symmetries], the phase diagram reduces to that of the one-fluid IBM: a point of second order phase transition between U(5) and SO(6) is embedded in a curve of first order transition between spherical and deformed equilibria.[28, 29] When we consider the full, three-dimensional phase diagram, the point of second order transition extends to form a *line* of second order transition, and the curve of first order transition extends to form a *surface* of first order transition.

Rather than considering the spectrosopy throughout the whole parameter space, let us here restrict our attention to the evolution of a few observables along the single line between $\mathrm{SU}_{\pi\nu}(3)$ and $\mathrm{SU}^*_{\pi\nu}(3)$. At the point of second order transition along this line, the classical equilibrium is just beginning to deviate from axial symmetry, and the spectroscopic properties [Fig. 4 (left)] still closely resemble those of a symmetric rotor. The $E2$ transition strengths between bands still very nearly follow the rotational Alaga rules. However, along the $\mathrm{SU}_{\pi\nu}(3)$–$\mathrm{SU}^*_{\pi\nu}(3)$ transition, the γ band rapidly plummets in energy, to become the degenerate $K = 2$ band of the $\mathrm{SU}^*_{\pi\nu}(3)$ ground state representation, and the $K = 4$ two-phonon γ excitation similarly descends to become the $K = 4$ band of the ground state representation. At the transition point, this leads to a substantial positive anharmonicity of the $K = 0$ two-phonon γ excitation. Beyond the transition point, $\mathrm{SU}^*_{\pi\nu}(3)$ triaxial features become more solidly apparent [Fig. 4 (right)], as the $E2$ branching ratios start to reflect the $\mathrm{SU}^*_{\pi\nu}(3)$ selection rules.

4. Symmetry energy

An essential aspect has so far been omitted from the analysis. The IBM-2 Hamiltonian is known to include a significant contribution from the Majorana operator, $\hat{M} \equiv -2\sum_{k=1,3}(d^\dagger_\pi \times d^\dagger_\nu)^{(k)} \cdot (\tilde{d}_\pi \times \tilde{d}_\nu)^{(k)} + (s^\dagger_\pi \times d^\dagger_\nu - s^\dagger_\nu \times d^\dagger_\pi)^{(2)} \cdot (\tilde{s}_\pi \times \tilde{d}_\nu - \tilde{s}_\nu \times \tilde{d}_\pi)^{(2)}$, which represents a proton-neutron symmetry energy. The geometric interpretation of this operator is that it is proportional to the squared difference between the proton and neutron deformation tensors [$\hat{M} \propto (\alpha_\pi - \alpha_\nu)^2$]. It thus energetically penalizes proton-neutron asymmetric deformations, of which the $\mathrm{SU}^*_{\pi\nu}(3)$ configuration [Fig. 2 (left)] is an extreme example.

The effect of the Majorana operator on the equilibrium configuration is to move the proton and neutron equilibrium coordinate values closer together, shifting equilibrium configurations which are proton-neutron triaxial (*e.g.*, $\gamma_\pi = 0°$ and $\gamma_\nu = 60°$) towards those which are one-fluid triaxial (*e.g.*, $\gamma_\pi \approx \gamma_\nu \approx 30°$). Further inspection shows that the Majorana operator makes the energy surface soft in γ about this triaxial minimum. A structural evolution is thus expected from proton-neutron triaxiality, through one-fluid triaxiality, to one-fluid γ-softness

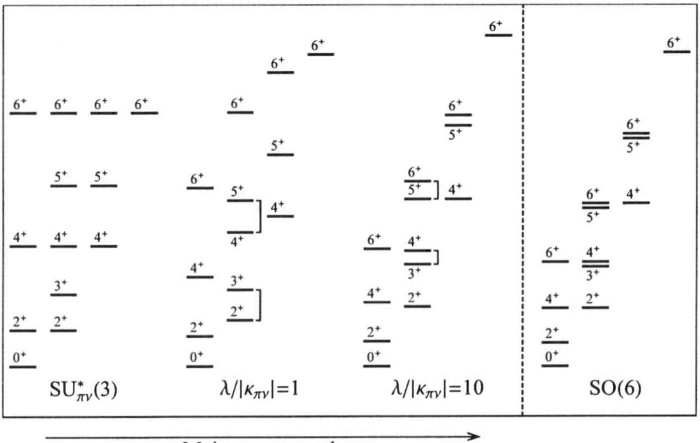

Figure 5. Level schemes showing the effect of introducing a Majorana term to the $\mathrm{SU}^*_{\pi\nu}(3)$ dynamical symmetry Hamiltonian. Note the evolution from $\mathrm{SU}^*_{\pi\nu}(3)$ rotational energy spacings, though Davydov-like energy staggering, to $\mathrm{SO}(6)$-like staggering. Level energies for the $\mathrm{SO}(6)$ symmetry are shown for comparison.

with increasing Majorana strength. This evolution is indeed evident in the spectroscopy (Fig. 5). A small Majorana contribution in the Hamiltonian results in a $(2^+3^+)(4^+5^+)\ldots$ staggering of level energies, like that of the Davydov model[24] of one-fluid triaxiality. A larger Majorana contribution leads to the reverse $2^+(3^+4^+)(5^+6^+)\ldots$ staggering, characteristic of γ-soft structure. Furthermore, the $E2$ transition strengths evolve towards those of $\mathrm{SO}(6)$, and the $M1$ admixtures become highly suppressed.

The coefficient λ of the Majorana operator in the Hamiltonian can be deduced from the energies of mixed symmetry or scissors excitations,[30] suggesting $\lambda/|\kappa_{\pi\nu}| \approx 5$ to be a reasonable estimate for nuclei near stability. Such a Majorana strength is found to severely limit the portion of the phase diagram in which triaxiality occurs and to largely wash out the spectroscopic signatures of proton-neutron triaxial structure [Fig. 5]. However, for nuclei far from stability, in which valence protons and neutrons occupy well-separated orbitals, proton-neutron interaction strengths can be expected to be reduced. Even a moderate reduction of the Majorana interaction strength (by a factor of 2 to 4) relative to the proton-proton and neutron-neutron quadrupole interactions yields a situation much more conducive to true proton-neutron triaxial structure.

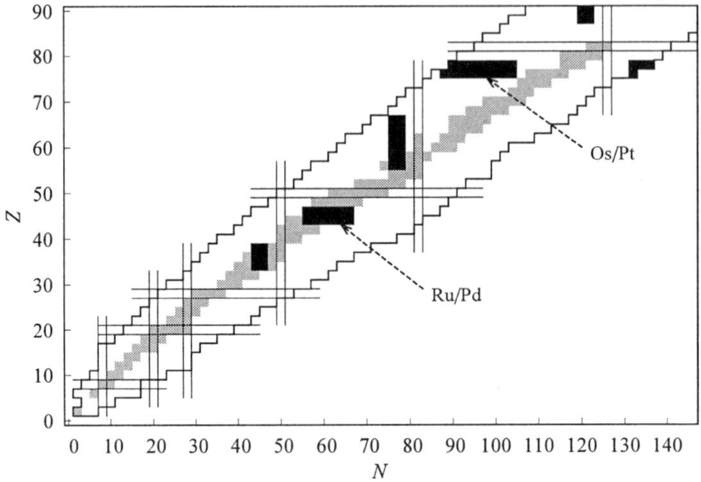

Figure 6. Candidate regions for proton-neutron triaxial structure. Regions below mid-shell for protons and just before the end of the shell for neutrons or *vice versa* are indicated in black (see text). Light gray squares indicate stable nuclides, double lines mark shell closures, and the larger outlined region contains nuclei accessible for spectroscopy at a next-generation radioactive beam facility. Figure adapted from Ref. 21.

5. Conditions for asymmetry

The collective structure of nuclei depends upon many aspects of the underlying single-particle structure: the energy spacings between orbitals (subshell gaps), the ordering of orbitals (low j or high j), and, especially for exotic nuclei, the radial wave functions (compact or diffuse). These single-particle properties are manifested in the effective interactions of the collective model. In the IBM-2, the dominant interactions include the pairing interactions (s-wave, d-wave, ...), the mutipole interactions (quadrupole, ...), and the symmetry energy (or Majorana operator).

For proton-neutron triaxiality to occur, χ_π and χ_ν must be large and of opposite sign. The classic "rule of thumb" is that χ is negative for particle-like valence nucleons, *i.e.*, near the beginning of a shell, and positive for hole-like valence nucleons, *i.e.*, near the end of a shell. In actuality, the relationship between shell filling and χ is much more complex.[31,32] However, assuming this simple relation and taking the shell closures valid near stability provides the candidate regions of Fig. 6.

6. Conclusion

In preparation for the study of nuclei at an exotic beam facility, proton-neutron asymmetric collective structure has been investigated within the framework of the IBM-2. Proton-neutron asymmetry is suppressed by the Majorana interaction near stability, but it could play a role for nuclei far from stability. The $SU^*_{\pi\nu}(3)$ dynamical symmetry is an idea limit, not likely to be reached in its pure form, but it illustrates the basic characteristics of proton-neutron triaxiality. The full analysis involves determination of the phase diagram, investigation of the nature of the phase transitions, and characterization of the signatures of asymmetric structure.

A natural extension of this work is to consider the Bose-Fermi systems (odd mass or odd-odd nuclei) consisting of a bosonic core and unpaired nucleons. This problem is of interest since odd nuclei will play a major role in shell structure studies and since coupling to an unpaired nucleon significantly influences the collective structure of the even-even core (core polarization). An appropriate framework for the analysis is the interacting boson fermion model[33, 34] (IBFM).

Acknowledgements

This work was supported by the US DOE under grant DE-FG02-91ER-40608.

References

1. J. Dobaczewski, I. Hamamoto, W. Nazarewicz, and J. A. Sheikh, *Phys. Rev. Lett.* **72**, 981 (1994).
2. N. Lo Iudice and F. Palumbo, *Phys. Rev. Lett.* **41**, 1532 (1978).
3. F. Iachello, *Nucl. Phys.* A **358**, 89c (1981).
4. D. Bohle, A. Richter, W. Steffen, A. E. L. Dieperink, N. Lo Iudice, F. Palumbo, and O. Scholten, *Phys. Lett.* B **137**, 27 (1984).
5. F. Iachello, *Phys. Rev. Lett.* **53**, 1427 (1984).
6. N. Pietralla, C. Fransen, P. von Brentano, A. Dewald, A. Fitzler, C. Frießner, and J. Gableske, *Phys. Rev. Lett.* **84**, 3775 (2000).
7. M. Goldhaber and E. Teller, *Phys. Rev.* **74**, 1046 (1948).
8. A. Zilges, S. Volz, M. Babilon, T. Hartmann, P. Mohr, and K. Vogt, *Phys. Lett.* B **542**, 43 (2002).
9. S. Frauendorf, *Rev. Mod. Phys.* **73**, 463 (2001).
10. P. Van Isacker and Jin-Quan Chen, *Phys. Rev.* C **24**, 684 (1981).
11. K. Heyde, P. Van Isacker, M. Waroquier, G. Wenes, Y. Gigase, and J. Stachel, *Nucl. Phys.* A **398**, 235 (1983).
12. F. Iachello and A. Arima, *The Interacting Boson Model* (Cambridge University Press, Cambridge, 1987).
13. A. Arima, T. Otsuka, F. Iachello, and I. Talmi, *Phys. Lett.* B **66**, 205 (1977).
14. P. Van Isacker, K. Heyde, J. Jolie, and A. Sevrin, *Ann. Phys. (N.Y.)* **171**, 253 (1986).
15. A. E. L. Dieperink and R. Bijker, *Phys. Lett.* B **116**, 77 (1982).

16. A. E. L. Dieperink and I. Talmi, *Phys. Lett.* B **131**, 1 (1983).
17. N. R. Walet and P. J. Brussaard, *Nucl. Phys.* A **474**, 61 (1987).
18. A. Sevrin, K. Heyde, and J. Jolie, *Phys. Rev.* C **36**, 2621 (1987).
19. A. Sevrin, K. Heyde, and J. Jolie, *Phys. Rev. C* **36**, 2631 (1987).
20. M. A. Caprio and F. Iachello, *Ann. Phys. (N.Y.)* **318**, 454 (2005).
21. M. A. Caprio, in *Nuclei and Mesoscopic Physics*, ed. V. Zelevinsky, AIP Conf. Proc. No. 777 (AIP, Melville, New York, 2005), p. 199.
22. A. Leviatan and M. W. Kirson, *Ann. Phys. (N.Y.)* **201**, 13 (1990).
23. T. Otsuka, *Hyperfine Interact.* **75**, 23 (1992).
24. A. S. Davydov and G. F. Filippov, *Nucl. Phys.* **8**, 237 (1958).
25. M. A. Caprio and F. Iachello, *Phys. Rev. Lett.* **93**, 242502 (2004).
26. J. M. Arias, J. Dukelsky, and J. E. García-Ramos, *Phys. Rev. Lett.* **93**, 212501 (2004).
27. J. N. Ginocchio and A. Leviatan, *Ann. Phys. (N.Y.)* **216**, 152 (1992).
28. A. E. L. Dieperink, O. Scholten, and F. Iachello, *Phys. Rev. Lett.* **44**, 1747 (1980).
29. D. H. Feng, R. Gilmore, and S. R. Deans, *Phys. Rev.* C **23**, 1254 (1981).
30. O. Scholten, K. Heyde, P. Van Isacker, J. Jolie, J. Moreau, M. Waroquier, and J. Sau, *Nucl. Phys.* A **438**, 41 (1985).
31. A. Van Egmond and K. Allaart, *Nucl. Phys.* A **425**, 275 (1984).
32. T. Otsuka, *Nucl. Phys.* A' **557**, 531c (1993).
33. F. Iachello and P. Van Isacker, *The Interacting Boson-Fermion Model* (Cambridge University Press, Cambridge, 1991).
34. R. Bijker, in these proceedings.

NO-CORE SHELL MODEL FOR NUCLEAR STRUCTURE AND REACTIONS

B. R. BARRETT, S. QUAGLIONI AND I. STETCU
Department of Physics, University of Arizona,
Tucson, AZ 85721-0081, USA
E-mail: bbarrett@physics.arizona.edu

P. NAVRÀTIL AND W. E. ORMAND
Lawrence Livermore National Laboratory, P.O. Box 808,
Livermore, CA 94551, USA

J. P. VARY
Department of Physics and Astronomy, Iowa State University,
Ames, IA 50011, USA

A. NOGGA
Institut für Kernphysik, Forshungszentrum Jülich,
52425 Jülich, Germany

The no-core shell model (NCSM), in which all A nucleons are treated as active, is an *ab initio* approach for calculating the properties of nuclei microscopically. We first present the basic NCSM formalism, which allows us to exactly determine the effective Hamiltonian as well as any general operator in a model space of arbitrary size for performing shell-model calculations. Good results for binding energies and spectra have been obtained for nuclei up to mass $A = 14$ using the best available two-nucleon (NN) and theoretical three-nucleon forces (3NF's). We next present three new applications and extensions of the NCSM approach: *1)* the use of the NCSM in the three-nucleon (3N) cluster approximation as a method for testing the reliability of different 3NF's, and the application of the NCSM to *2)* other general operators and *3)* inclusive electromagnetic responses.

1. Introduction

High precision nuclear experiments, which can shed light on various fundamental questions of nature, rely heavily on a good understanding of the underlying physics. Thus, interpretation of double-β decay experiments that are essential to pin down the Majorana/Dirac character of neutrinos[1] depends upon reliable understanding of the nuclear physics involved in description of the decay rate.[2] In astrophysics, it is widely accepted that the origin of a large number of isotopes is the r-process nucleosynthesis,[3] a phenomenon that creates elements heavier than iron by rapid neutron capture and involves neutron rich nuclei far from the valley of stability. In the absence of experimental data, current theoretical models, which describe correctly stable nuclei, tend to disagree widely in the region of interest for the r process. The proposed reaccelerated exotic beam facility (REBF) will offer invaluable guidance in the construction and validation of nuclear models, but will not measure all the nuclei of interest. This motivates the need for a reliable microscopic nuclear theory.

The NCSM provides a solution to the nuclear many-body problem for light nuclei. It is based on an effective Hamiltonian derived from realistic "bare" NN and 3N interactions acting within a finite Hilbert space. All A-nucleons are allowed to interact. The main tool is a unitary transformation, which takes into account correlations left out by the truncation of the Hilbert space. The high accuracy of the NCSM allows for the investigation of the reliability of the chiral nuclear interaction. This follows from the fact that the properties, e.g., energy spectra, of p-shell nuclei are sensitive to the subleading parts of the chiral interactions, including 3N forces.[4] Further investigations, of interest for applications to nuclei far from stability, should involve tests of other approaches, e.g., mean field and beyond, against solutions obtained with the NCSM. The NCSM is well suited to provide benchmarks for p-shell and sd-shell nuclei, where the latter methods can be applied, so that their accuracy can be checked.

2. No-core shell-model approach

We start from the intrinsic Hamiltonian for the A-nucleon system, i.e.,

$$H_A = \frac{1}{A}\sum_{i>j}\frac{(\vec{p}_i - \vec{p}_j)^2}{2m} + \sum_{i>j} V_{ij}^{NN}\,, \tag{1}$$

where m is the nucleon mass, and V_{ij}^{NN} the bare NN interaction, such as the Argonne potentials in coordinate space[5] or the non-local CD-Bonn potential.[6] In this paper, we do not present the generalization for the inclusion of three body forces, a detailed description of which can be found in Refs. 4, 7 and 8.

In the next step, we modify Eq. (1) by adding to it the harmonic oscillator (HO) center-of-mass Hamiltonian $\vec{P}^2/(2Am) + Am\Omega^2\vec{R}^2/2$,

$$\vec{P} = \sum_{i=1}^{A}\vec{p}_i \quad \text{and} \quad \vec{R} = \frac{1}{A}\sum_{i=1}^{A}\vec{r}_i\,. \tag{2}$$

The effect of the center-of-mass HO term will be subtracted in the final many-body calculation and does not influence the intrinsic properties of the many-body system. However, it does provide a single-particle basis for performing numerical calculations and binding of the nucleon cluster, important for the cluster approximation to the effective interaction discussed below, and improves the convergence of the many-body problem. The modified Hamiltonian, which possesses a pseudo-dependence on the HO frequency Ω, is given by:

$$\begin{aligned} H_A^\Omega &= H_A + \frac{\vec{P}^2}{2mA} + \frac{1}{2}mA\Omega^2R^2 \\ &= \sum_{i=1}^{A}\left[\frac{\vec{p}_i^2}{2m} + \frac{1}{2}m\Omega^2\vec{r}_i^2\right] + \sum_{i<j=1}^{A}\left[V_{ij}^{NN} - \frac{m\Omega^2}{2A}(\vec{r}_i - \vec{r}_j)^2\right] . \end{aligned} \tag{3}$$

Since we use a finite model space in order to solve the many-body problem, we employ a unitary transformation[9,10] to accommodate the short-range two-body correlations:

$$\mathcal{H} = e^{-S}H_A^\Omega e^S . \tag{4}$$

The infinite Hilbert space associated with the system is split into the finite model, or P-space, and the complementary, or Q-space, with the projectors P and Q spanning the entire space, $P + Q = 1$. The model space P is defined by all basis states up to an energy of $N_{max}\hbar\Omega$ above the ground-state configuration, where N_{max} is an integer. The transformation S is determined, so that one obtains energy-independent effective operators, and is equivalent with the decoupling conditions $P\mathcal{H}Q = 0$ and $Q\mathcal{H}P = 0$. Analogously to Eq. (4), general operators are also transformed by the same transformation:

$$\mathcal{O} = e^{-S}Oe^S. \tag{5}$$

The operator S, with the restrictions $PSP = QSQ = 0$, can be formally written[11] by means of another operator ω as

$$S = \text{arctanh}(\omega - \omega^\dagger), \tag{6}$$

with $Q\omega P = \omega$. After some algebra, one obtains the following expressions for the energy-independent effective Hamiltonian and an arbitrary effective operator[10] in the P space, respectively:

$$H_{eff} = P\mathcal{H}P = \frac{P + P\omega^\dagger Q}{\sqrt{P + \omega^\dagger\omega}}H\frac{P + Q\omega P}{\sqrt{P + \omega^\dagger\omega}} , \tag{7}$$

$$O_{eff} = P\mathcal{O}P = \frac{P + P\omega^\dagger Q}{\sqrt{P + \omega^\dagger\omega}}O\frac{P + Q\omega P}{\sqrt{P + \omega^\dagger\omega}} . \tag{8}$$

A very efficient way to compute the operator ω is[12]

$$\langle \alpha_Q|\omega|\alpha_P\rangle = \sum_k \langle \alpha_Q|k\rangle\langle \tilde{k}|\alpha_P\rangle, \tag{9}$$

where $|\alpha_P\rangle$ and $|\alpha_Q\rangle$ are the basis states of the P and Q spaces, respectively; $|k\rangle$ denotes states from a selected set $\mathcal{K}$ of exact eigenvectors of the Hamiltonian H in the full A-body space, and $\langle \tilde{k}|\alpha_P\rangle$ is the matrix element of the inverse overlap matrix $\langle \alpha_P|k\rangle$. Hence, computing ω by means of Eq. (9) yields the exact solution of the initial A-body problem, which is our goal. For practical applications we approximate the effective interaction as a sum of a-body effective interaction terms ($a < A$), derived from the solution of the a-body problem. This approximation, known as the cluster approximation, leads to the exact solution in the limit $a \to A$ and the contribution from higher-order clusters becomes less and less important in the limit when the model space is very large. For further details, we refer the reader to previous publications, e.g.,[13] and references therein.

3. Calculation of energy spectra for light nuclei

Most of the NCSM applications to light nuclei have been to the description of the energy spectrum, from ^{3}H to ^{16}O.[8,13–16] Among the *ab initio* methods, the NCSM is the most flexible, allowing treatment of both local and non-local NN interactions, as well as a much larger mass range, with an estimated limit of applicability around mass 24.

A uniform feature of all NCSM calculations with only NN interactions, as well as for other microscopic approaches for calculating the spectra of light nuclei,[17] is that they consistently underbind ground states of nuclei and do not converge to the excited state energies.[18,19] The general consensus is that one needs additional 3NF's, or perhaps even more-nucleon forces, to explain these discrepancies. This seems like a reasonable assumption, because three- and more-nucleon forces naturally arise in the determination of the nuclear force from effective field theory using a chiral perturbation theory (CPT) approach.[20,21]

At the present time *ad hoc* combinations of NN and 3NF's can provide the correct binding energies for the three- and four-nucleon systems,[22,23] but some fail for p-shell nuclei and some 3N scattering observables. Although, the application of CPT provides consistent NN and 3NF models,[20,21] this approach yields so-called contact terms, *i.e.*, constant terms, which must presently be determined by fitting experimental data. In the CPT approach 3NF's first appear at next-to-next-lowest order (N2LO) and have two contact terms, with two strength constants called c_D and c_E. When these two strength constants are determined for the Idaho-N3LO interaction,[24] by fitting to the binding of ^{3}H and ^{3}He,[4] we find

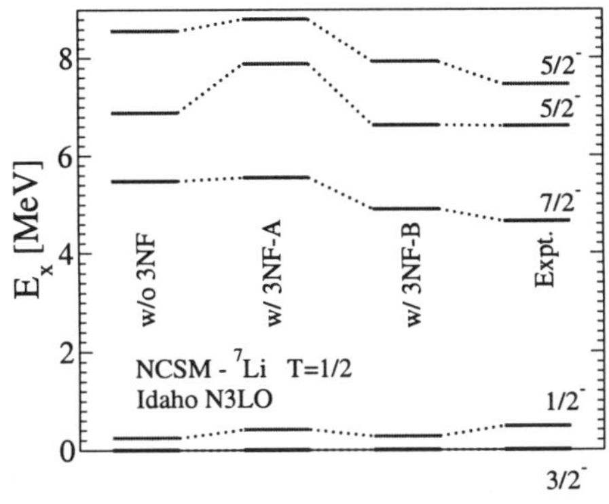

Figure 1. Dependence of the excitation energy of the lowest states of ^{7}Li on the interaction. Results with the NN interaction only and with the 3NF-A and 3NF-B included are compared to the experimental value.

that two sets of values , called 3NF-A and 3NF-B, can be found, which determine these data exactly.

We note that the NCSM offers a way to test these different theoretical 3NF's by performing nuclear-structure calculations at the three-body cluster level for p-shell nuclei. Equation (9) for determining the matrix elements of the transformation operator ω is applicable at any cluster level, *i.e.*, for any $a \leq A$. Thus, if we now insert a 3NF into Eqs. (1) and (3) and perform the unitary transformation (4) at the three-body cluster level, *i.e.*, $a = 3$, we can obtain the three-body matrix elements of ω (9) and, thereby, the three-body matrix elements of H_{eff} (7), which now contain contributions from both the NN interaction and the theoretical 3NF. One can determine the effect of adding the theoretical 3NF by performing the three-body cluster calculation first with only the NN interaction and then, a second time, including the 3NF.

We have performed the first such calculations for ^{6}Li[25] and ^{7}Li,[4] so as to test the feasibility of this concept. Details of this approach along with numerous results are given in Ref. 4. In Fig. 1 we show the results for the low-energy spectrum of ^{7}Li, which are marginally better for the choice 3NF-B over 3NF-A. Clearly, more calculations for other p-shell nuclei need to be carried out before a more definitive conclusion can be drawn.

However, it seems fairly obvious at this time that one possibility for determining the unknown terms in theoretical 3NF's will come from the comparison of results of nuclear-structure calculations with experimental spectra of p-shell nuclei. In this regard, the NCSM offers a straightforward method for performing these calculations and making these comparisons.

4. Other effective operators

Application of the unitary transformation to general operators is much more involved than for the Hamiltonian. Moreover, for long-range operators, such as the quadrupole transition operator, it has been found that the renormalization in the two-body cluster approximation has little effect.[26] For example, in the case of the quadrupole transition $1^+0 \rightarrow 3^+0$ for ^{6}Li, in the $2\hbar\Omega$ model space, one obtains $B(E2) = 2.647\ e^2 fm^4$ using the bare operator, and a negligible renormalization when the corresponding effective operator is employed, i.e., $B(E2) = 2.784 e^2 fm^4$.[26] However, in $10\hbar\Omega$, using the bare operator, one obtains $B(E2) = 10.221\ e^2 fm^4$. In principle, since one expects larger renormalization in the smaller model space (one has to take into account a larger excluded space), the value obtained with the effective operator in the $2\hbar\Omega$ model space should be significantly closer to the result in $10\hbar\Omega$. Since this is obviously not the case, one concludes that in the two-body cluster approximation the effect of the renormalization for long-range operators is very small. This has been demonstrated for a two-body Gaussian operator of variable range[26] and for the longitudinal-longitudinal distribution function.[27] These calculations show that short-range operators can be reliably calculated in the two-body cluster approximation, even in small spaces, when using the appropriate effective operator.[27] On the other hand, long-range operators are only weakly renormalized at the two-body cluster level. In order to accommodate the long-range correlations, one has to increase the model space and/or use a higher-order cluster approximation.

5. Inclusive electromagnetic responses

In the inclusive cross sections of reactions induced by perturbative external probes all the relevant information on the dynamics of the nuclear target is contained in the so-called response functions defined as:

$$R(\omega) = \int d\Psi_f \left| \langle \Psi_f | \hat{O} | \Psi_0 \rangle \right|^2 \delta(E_f - E_0 - \omega) \,, \tag{10}$$

where ω represents the energy transferred by the probe, and $\hat{O}$, the excitation operator. Wave functions and energies of the ground and final states of the perturbed system are denoted by $|\Psi_{0/f}\rangle$ and $E_{0/f}$, respectively.

In light systems the discrete spectra are very limited. Therefore, the final states $|\Psi_f\rangle$ in the continuum already lie at low energy. Because of the enormous difficulties in calculating many-body scattering states for $A > 3$, this represents a major obstacle, if one wants to calculate responses in the continuum directly from Eq. (10). Such difficulties can be avoided using the Lorentz integral transform

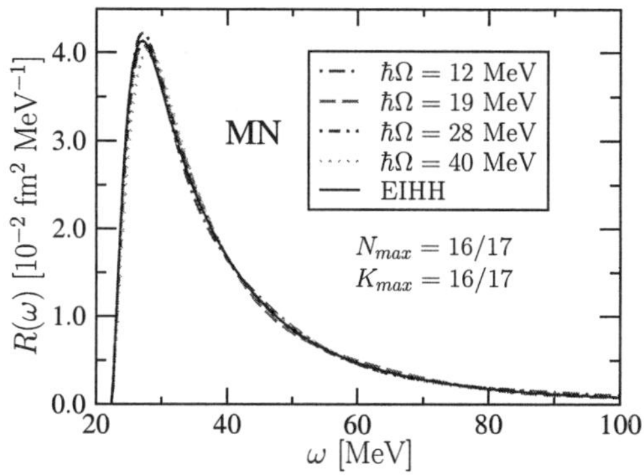

Figure 2. The NCSM ($N_{max} = 16/17$) inclusive response to the isovector dipole transition [see Eq. (10)] as a function of ω for four different values of the HO parameter Ω: comparison with the EIHH ($K_{max} = 16/17$) result.

(LIT) approach[28], which reduces the continuum problem to that for a bound-state-like problem, so that only bound state techniques are required.

In the LIT method one obtains $R(\omega)$ after the inversion of an integral transform with a Lorentzian kernel

$$L(\sigma_R, \sigma_I) = \int d\omega \frac{R(\omega)}{(\omega - \sigma_R)^2 + \sigma_I^2} = \langle \widetilde{\Psi} | \widetilde{\Psi} \rangle \,. \tag{11}$$

The state $\widetilde{\Psi}$ is the unique solution of the inhomogeneous "Schrödinger-like" equation

$$(H - E_0 - \sigma_R + i\sigma_I)|\widetilde{\Psi}\rangle = \hat{O}|\Psi_0\rangle. \tag{12}$$

Because of the presence of an imaginary part σ_I in Eq. (12) and the fact that the right-hand side of this same equation is localized, one has an asymptotic boundary condition similar to a bound state. Thus, one can apply bound-state techniques for its solution, and, in particular, expansions over basis sets of localized functions.

Recently we have examined the applicability of the NCSM approach to the solution of Eq. (12) by performing a test consisting in calculating, within the effective interaction hyperspherical harmonics (EIHH)[29] and NCSM approaches, the LIT of the four-body response functions to two different excitation operators.[30] The input interaction was the simple Minnesota (MN)[31] potential, which was more convenient for the purposes of our comparison.

As an example, in Fig. 2 we present the EIHH and NCSM results for the ^{4}He inclusive response to the isovector dipole excitation, obtained by means of the LIT method.

The NCSM calculations are carried out for four different HO frequencies ($\hbar\Omega = 12, 19, 28$ and 40 MeV), in order to study the dependence of the resulting response on this parameter and to provide an estimate for the theoretical uncertainties. All the curves, excluding the result for $\hbar\Omega = 40$ MeV, not yet completely converged, show a good agreement. The HO frequency value of $\hbar\Omega = 19$ MeV leads to the best agreement with the EIHH response. Indeed, the discrepancy between the two numerical results does not exceed 5% in the energy intervals immediately close to the disintegration threshold as well as in the dipole resonance peak region. While the resonant peak is equally well described for $\hbar\Omega = 12$ and 28 MeV, these HO frequencies yield a larger discrepancy (at most 10% and 15%, respectively) within 3 MeV from threshold. The poorest agreement is found for $\hbar\Omega = 40$ MeV, although mainly in the low-energy part of the response. In the range 60 MeV$\leq \omega \leq$ 80 MeV all four NCSM responses agree within the 7% or better of the EIHH result.

The level of precision reached by the NCSM in this benchmarking calculation with a semirealistic NN interaction is encouraging in the perspective of LIT investigations on heavier nuclei. Indeed, the NCSM has the ability of handling heavier mass nuclei than the EIHH. On the other hand, the large model spaces needed in the NCSM calculations in order to achieve such accuracy suggest that a more substantial numerical effort will be necessary.

6. Summary

The goal of nuclear theory is an exact treatment of nuclei based on NN and 3N interactions. Techniques for calculating the structure of light nuclei, such as the NCSM, are able to achieve this goal. However, we still need a bridge between our *ab initio* few-body and light-nuclei calculations, *i.e.*, $A \lesssim 24$, and $0\hbar\Omega$ standard shell-model calculations ($16 \lesssim A \lesssim 60$) and density-functional-theory calculations ($A \gtrsim 60$). In this regard, the NCSM will serve in a supporting role to the REBF by providing the benchmark calculations for input to these studies of heavier nuclei.

By investigating the intersections between these theoretical strategies, *i.e.*, by probing the structure of medium and heavy nuclei off the line of stability, the REBF will provide the experimental tool for developing the unified description of the nucleus. At the same time, the NCSM contributes in two ways to more predictive calculations for heavier nuclei. First, it can provide essential benchmarks for p-shell and light sd-shell nuclei to improve and check less exact methods. Second, it enables the study of nuclear interactions based on CPT, which connect properties of nuclear interactions to the symmetries of QCD. This is essential in order to understand, e.g., the charge dependence of NN and 3N forces.

Acknowledgments

We thank S. Bacca, N. Barnea, W. Leidemann and G. Orlandini for their contribution to part of the work in this paper (EIHH calculations) and C. W. Johnson for helpful discussions. B.R.B, S.Q. and I.S. acknowledge partial support by NFS grants PHY0070858 and PHY0244389. The work was performed in part under the auspices of the US DOE under grant No. DE-FC02-01ER41187, DEFG02000ER41132, DE-FG-02-87ER-0371, and DE-AC02-76SF00515 and Lawrence Livermore National Laboratory under contract No. W-7405-Eng-48. P.N. received support from LDRD contract 04- ERD-058.

References

1. S. Elliot and P.Vogel, *Annu. Rev. Nucl. Part. Sci.* **52**, 115 (2002).
2. J. Engel and P. Vogel, *Phys. Rev.* **C69**, 034304 (2004).
3. D. D. Clayton, *Principles of Stellar Evolution and Nucleosynthesis*, The University of Chicago Press, Chicago, 1983.
4. A. Nogga, P. Navrátil, B. R. Barrett and J. P. Vary, *Preprint* nucl-th:0511082 (2005), to be published in *Phys. Rev.* **C**.
5. R. B. Wiringa, V. G. J. Stocks and R. Schiavilla, *Phys. Rev.*, **C51**, 38 (1995); S. Pieper and R. B. Wiringa, *Annu. Rev. Nucl. Part. Sci.* **51**, 53 (2001).
6. R. Machleidt, F. Sammarruca and Y. Song, *Phys. Rev.* **C53**,1483 (1996).
7. P. Navrátil and W. E. Ormand, *Phys. Rev. Lett.* **88**, 152502 (2002).
8. P. Navrátil and W. E. Ormand, *Phys. Rev.* **C68**, 034305 (2003).
9. K. Suzuki and S. Y. Lee, *Prog. Theor. Phys.* **64**, 2091 (1980); K. Suzuki, *op. cit.* **68**, 246 (1982); S. Okubo, *op. cit.* **12**, 603 (1954).
10. P. Navrátil, H. Geyer and T. T. S. Kuo, *Phys. Lett.* **B315**, 1 (1993).
11. K. Suzuki, *Prog. Theor. Phys.* **68**, 1999 (1982); K. Suzuki and R. Okamoto, *op. cit.* **92**, 1045 (1994).
12. P. Navrátil and B. R. Barrett, *Phys. Rev.* **C54**, 2986 (1996).
13. P. Navrátil, J. P. Vary and B. R. Barrett, *Phys. Rev.* **C62**, 054311 (2000).
14. P. Navrátil and B. R. Barrett, *Phys. Rev.* **C57**, 562 (1998); P. Navrátil, G. P. Kamuntavičius and B. R. Barrett, *op. cit.* **C61**, 044001 (2000); P. Navrátil, J. P. Vary and W. E. Ormand, *Phys. Rev. Lett.* **87**, 172502 (2001).
15. S. Aroua, P. Navrátil, L. Zamick, M. S. Fayache, B. R. Barrett, J. P. Vary, N. Smirnova and K. Heyde, *Nucl. Phys.* **A720**, 71 (2003).
16. J. P. Vary *et al.*, *Eur. Phys. J* **A25**, Suppl. 1, 475 (2005).
17. S. C. Pieper, K. Varga, R. B. Wiringa and J. Carlson, *Phys. Rev.* **C66**, 044310 (2002); S. C. Pieper, R. B. Wiringa and J. Carlson, *op. cit.* **C70**, 054325 (2004).
18. H. Kamada *et al.*, *Phys. Rev.* **C64**, 044001 (2001).
19. P. Navrátil and E. Caurier, *Phys. Rev.* **C69**, 014311 (2004).
20. P. Bedaque and U. van Kolck, *Annu. Rev.Nucl. Part. Sci.* **52**, 339 (2002).

21. E. Epelbaum, *Preprint* nucl-th:0509032 (2005), to appear in *Prog. Part. Nucl. Phys.*.
22. A. Nogga, H. Kamada, W. Glöckle and B. R. Barrett, *Phys. Rev.* **C65**, 054003 (2002).
23. S. C. Pieper, V. R. Pandharipande, R. B. Wiringa and J. Carlson, *Phys. Rev.* **C64**, 014001 (2001);
24. D. R. Entem and R. Machleidt, *Phys. Rev.* **C68**, 041001(R) (2003).
25. B. R. Barrett, P. Navrátil, A. Nogga, W. E. Ormand and J. P. Vary, *Nucl. Phys.* **A746**, 579 (2004).
26. I. Stetcu, B. R. Barrett, P. Navrátil, J. P. Vary, *Phys. Rev.* **C71**, 044325 (2005).
27. I. Stetcu, B. R. Barrett, P. Navrátil, J. P. Vary, *Phys. Rev.* **C73**, 037307 (2006).
28. V. D. Efros, W. Leidemann and G. Orlandini, *Phys. Lett.* **B338**, 130 (1994).
29. N. Barnea, W. Leidemann and G. Orlandini. *Phys. Rev.* **C61**, 054001 (2000); *op. cit.* **C67**, 054003 (2003).
30. I. Stetcu, S. Quaglioni, S. Bacca, B. R. Barrett, C. W. Johnson, P. Navrátil, N. Barnea, W. Leidemann and G. Orlandini, to be submitted.
31. D. R. Thomson, M. LeMere and Y. C. Tang, *Nucl. Phys.* **A286**, 53 (1977).

CLUSTERING IN NEUTRON-RICH NUCLEI

HISASHI HORIUCHI
Research Center for Nuclear Physics, Osaka University,
Ibaraki city, Osaka, 567-0047, Japan
**E-mail: horiuchi@rcnp.osaka-u.ac.jp*

Two kinds of clustering generated or supported by excess neutrons are discussed: one is the formation of α cluster structure and the other is the formation of di-neutron clusters.

1. Introduction

In neutron-rich nuclei we observe the formation of neutron skins and neutron halos where neutrons are weakly bound with dilute density. Studies of correlations and clustering which are due to weakly bound dilute neutrons are important subjects of nuclear structure.

In this talk, two kinds of clustering are discussed which are generated or supported by excess neutrons. They are a) α cluster structure and b) di-neutron clusters. As for the α clustering, multi-cluster structure is also briefly mentioned in addition to the two-cluster structure.

2. Neutrons around di-cluster core

The nucleus ^{8}Be is known to have α-α cluster structure, which has been verified by an *ab initio* calculation by Wiringer et al. [1] recently. The study of neutron-rich Be isotopes by AMD (antisymmetrized molecular dynamics) [2] showed that these isotopes also have the α-α cluster structure which is surrounded by excess neutrons. The excess neutrons occupy the molecular orbits around the α-α core. The molecular orbit which is the p-shell orbit in the zero deformation limit is the π-orbit. The important molecular orbit is the so-called σ-orbit coming down from the sd-shell in the zero deformation limit. A very distinctive feature of the Be isotopes is the fact that we have many observed data of the excited states in each Be isotope and the structures of the excited states are well described by the configurations of σ^0, σ^1, and σ^2. Here, σ^p means the configuration in which p neutrons occupy the σ- orbit.

The ground states of ^{9}Be and ^{10}Be have the σ^0 confiruration, but those of ^{11}Be and ^{12}Be have the σ^1 and σ^2 configurations, respectively. When neutrons occupy the σ-orbit, the deformation of the system, or the degree of clustering of the system, increases. It is because the system can avoid the increase of the kinetic energy due to the occupation of σ-orbit by increasing the distance between two α clusters. Therefore the clustering of ^{11}Be is larger than that of ^{10}Be, and that of ^{12}Be is larger than that of ^{11}Be. The σ^1 structure of ^{11}Be ground state gives rise to the spin-parity $1/2^+$, which is the well-known parity inversion of ^{11}Be.

In ^{9}Be we have the low-lying $1/2^+$ excited state which is well described by the σ^1 structure. In ^{10}Be we have the low-lying $K^\pi = 1^-$ and $K^\pi = 0^+$ excited rotational bands which are well described by the σ^1 and σ^2 structures, respectively. AMD calculations of ^{10}Be reproduce not only the excitation energies but also the B(E2) and B(GT) values. The deformation (inter-α distance) of the σ^2 structure is larger than that of the σ^1 structure. In ^{12}Be, we have shell-model-like 0^+ state below 5 MeV and "^{6}He-^{6}He"-like molecular band above 10 MeV. These features are well reproduced by AMD calculations [2].

As is seen in the above explanations of Be isotopes, the formation of the α cluster structure in neutron-rich Be isotopes is largely due to the action of the excess neutrons which occupy the σ-orbits.

Recently M.Kimura [3] has discussed a possibility of the formation of the neutron molecular orbits around α-^{16}O core in ^{22}Ne. According to his study, while the ground band has a mean-field-like structure, the σ^2 molecular bands appear with the band-head states around 8 MeV, and the α-^{18}O band appear above 15 MeV.

3. Di-neutron cluster

P.G. Hansen discussed the neutron halo of ^{11}Li by using the di-neutron cluster model, ^{11}Li = ^{9}Li + di-neutron. G. Bertsch and H. Esbensen solved 3-body model of ^{9}Li + n + n and found the formation of di-neutron in the surface region [4]. However, experimental values of B(E1) strength distributions obtained by Coulomb breakup reactions were not in good agreements with Bertsch- Esbensen results. If ^{11}Li has ef^{9}Li + di-neutronffstructure, the peak position of the B(E1) strength distribution is expected to be around 0.47 MeV, however experimental data gave much higher position of the peak. Moreover, unfortunately, the discrepansies among the experimental results obtained at GSI, RIKEN, and MSU, made it difficult to persue in more detail theoretically the validity of the di-neutron cluster model. Recently T.Nakamura and his group remeasured the breakup reaction of ^{11}Li with improved experimental procedure at RIKEN, and obtained new results which differ from previous data from three institutions [5]. The peak position of the new data is now much closer to E(peak) $\sim$ 0.47 MeV. Furthermore, B(E1) distribution due to new data is now in good agreement with Bertsch-Esbensen results.

Here I present the results of a similar ^{9}Li + n+n 3-body calculation as Bertsch and Esbensen which was made by Hagino and Sagawa [6] by introducing the recoil correction term into the 3-body Hamiltonian. They of course could reproduce the Nakamura's data of B(E1) strength distribution. They analysed the two neutron density distribution for each of two-neutron spin $S = 0$ and 1. The analysis showed that for $S = 0$ the dineutron is clearly formed with its halo tail while for $S = 1$ there is no prominent density distribution for small opening angle between two neutrons. The sum of S=0 and S=1 density distribution is of course essentially the same as that of Bertsch and Esbensen.

Besides ^{11}Li, ^{6}He was also studied by Hagino and Sagawa. They also pointed the formation of a di-neutron cluster for $S = 0$ although the so-called cigar-shaped structure is recognizable unlike ^{11}Li. As for the di-neutron problem of ^{6}He, I refer to the work by Aoyama et al. [7] who pointed out following two important points. The first point is the fact that in order to have convergence of the shell-model type calculation we need the configuration mixing up to the very high angular-momentum orbits of valence neutrons. It is because the small opening angle between two neutrons requires the inclusion of the very high angular-momentum orbits of neutrons. The second point is the importance of the T-type coordinate system for the 3-body system of α+ $n + n$. The T-type coordinate system consists of the $n - n$ relative coordinate and the relative coordinate between α and $2n$. The rapid convergence of the calculation with the T-type coordinate system is just the reflection of the formation of the di-neutron cluster.

Recently Matsuo and his collaborators have reported the prominent di-neutron correlation obtained in the HFB (Hatree Fock Bogoliubov) calculations for medium-mass neutron-rich nuclei such as ^{22}O, ^{58}Ca, and ^{84}Ni [8]. Probabilty of two neutrons with anti-parallel spins to exist within short relative distance like 2 or 3 fm is very large especially in the low-density skin region. The configuration mixing of high-angular-momentum orbits is found, which is the necessary condition for the formatuin of strong di-neutron correlation.

4. Multi-cluster correlation

The possible existence of the linear α-chain structure has long been discussed although no clear experimental evidence has ever been obtained. In expecting the existence of the linear α-chain structure, it has been always discussed that if the system has excess neutrons the linear α-chain structure will be formed with more stability. Recently this problem was studied in detail in the case of neutron-rich C isotopes by Itagaki et al. [9]. They reported that if excess neutrons occupy suitable molecular orbits, a 3α-chain structure can be formed stably in ^{16}C.

Another example which shows that excess neutrons can stabilize an α-cluster structure is given by Ref.[10]. Here it is shown that the excess neutrons in ^{14}C stabilize the regular triangular shape of 3α clusters. The authors discuss that the calculated results of the excited $K^{\pi} = 3^{-}$ band are in good correspondence with the observed data.

5. Summarizing discussion

The first subject discussed here was the structure of 'di-cluster + neutrons'. It is pointed out that the excess neutrons are well described by molecular orbits and sometimes by atomic orbits. An important point is the fact that the formation of the α cluster structure in neutron-rich Be isotopes is largely due to the action of the excess neutrons which occupy the σ-orbits.

The second subject was about di-neutron cluster. Recent Coulomb breakup experiments of ^{11}Li indicate the existense of di-neutron cluster in ^{11}Li. ^{9}Li + n+n 3-body calculations well reproduce these data and clearly shows the formation of di-neutron cluster. It was reported that the prominent di-neutron correlation was obtained in the recent HFB calculations for medium-mass neutron-rich nuclei.

The third subject was about the multi-cluster structure stabilized by excess neutrons. Two works were discussed: one was triangle structure of 3 in ^{14}C and the other was linear chain structure of 3 in ^{16}C.

Formation of di-neutron cluster(s) in neutron-rich nuclei suggests the di-neutron condensation in neutron skin/halo of neutron-rich nuclei. The results of the recent HFB calculations support this idea. Similar phenomenon as the di-neutron condensation in dilute neutron matter is the cluster condensation in dilute excited states of finite nuclei. cluster condensation has been recently shown to exist in ^{12}C and has been strongly suggested to exist in ^{16}O. Future prospects of the study of cluster gas-like states are for the search of possibilities of the cluster condensation near-proton-dripline nuclei and in neutron-richer nuclei in medium-mass region.

Studies of correlations and clustering serve to elucidate the richness of nuclear dynamics. We expect new dynamics near driplines and also in excited states, We can expect various kinds of new dynamics based on mean-field dynamics, strong correlation dynamics, and clustering dynamics.

References

1. R.B.Wiringer, S.C.Pieper, J.Carlson, and V.R.Pandharipande, Phys. Rev. C **62**, 014001 (2000).
2. Y.Kanada-En'yo, M.Kimura, and H.Horiuchi, C.R. Physique **4**, 497 (2003).
3. M.Kimura, private communication.
4. G.F.Bertsch and H. Esbensen, Ann. Phys. (NY) **209**, 327 (1991); H. Esbensen and G.F.Bertsch, Nucl. Phys. A **542**, 310 (1992); H. Esbensen, G.F.Bertsch, and K.Hencken, Phys.Rev. C **56**, 3054 (1999).
5. T.Nakamura et al., RIKEN Accel. Prog. Rep. No.38.
6. K.Hagino and H.Sagawa, Phys. Rev. C **72**, 044321 (2005).
7. S.Aoyama, S.Mukai, K.Kato, and K.Ikeda, Prog. Theor. Phys. **93**, 99 (1995).
8. M.Matsuo, K.Mizuyama, and Y.Serizawa, Phys. Rev. C **71**, 064326 (2005).
9. N. Itagaki, S. Okabe, K.Ikeda, and I. Tanihata, Phys. Rev C **64**, 014301 (2001).
10. N. Itagaki, T.Otsuka, K.Ikeda, and S.Okabe, Phys. Rev. Letters **92**, 142501(2004).

QUANTUM MONTE CARLO: NOT JUST FOR ENERGY LEVELS

KENNETH M. NOLLETT
Physics Division, Argonne National Laboratory
9700 S. Cass Ave., Argonne, IL 60439, USA
E-mail: nollett@anl.gov

Tremendous progress has been made over the last dozen years in understanding light nuclei in terms of bare nucleon-nucleon interactions. In this contribution, I describe work based on the Argonne v_{18} two-body and Illinois three-body interactions, using the Variational Monte Carlo and Green's function Monte Carlo methods to compute wave functions and their energies. As presently implemented, these methods are limited to nuclei with $A \leq 12$, but much remains to be done even in this mass range. In the past, we have focused our effort on energies of bound and narrow states, computing some 68 levels and reproducing many features of the observed energy spectrum. We are now beginning to focus on other nuclear properties: both bound-state wave function properties like charge radii and beta decay rates, and scattering and reaction cross sections. I describe in particular detail our first *ab initio* scattering calculation and show computed cross sections for low-energy ^{4}He-n scattering.

1. Motivation

It is a fundamental goal of nuclear physics to understand atomic nuclei quantitatively and in the most microscopic way possible. For the foreseeable future, this means understanding them as collections of nucleons interacting via either their interactions in vacuum or effective interactions derivable from those interactions. The quantum Monte Carlo (QMC) methods provide a useful means toward this goal and have been the focus of efforts by a collaboration at Urbana, Argonne, Los Alamos, and elsewhere, including myself. The QMC methods differ from most other methods presently in use (see other contributions to this volume) in ways that present a distinct set of advantages and disadvantages.

The QMC methods work not with basis functions but with samples of the wave function at discrete points in the $3A$-dimensional configuration space. This removes all concern about convergence as a function of basis size or oscillator parameter. It also allows us to deal easily with interactions that have strong repulsive cores or three-body interactions. Since we do not have a truncated spatial basis, we do not need to construct or assume effective interactions, operators, or quenching. Without a spatial basis, neither intruder states nor unbound states require significantly more effort than natural-parity or bound states. We can treat an unbound or weakly-bound state as just another (vector) function of the $3A$ particle coordinates, without framing the problem in terms of "coupling to the continuum" or constructing an elaborate basis corresponding to that language.

These properties imply that our basis-choice and convergence uncertainties are very different from those of other nuclear many-body methods. The results of Green's function Monte Carlo are essentially exact, so they have obvious direct application to nuclei where they can be obtained. Even though QMC methods so far are restricted to light nuclei, they provide cross checks within their useful mass range for the more approximate methods that may be unavoidable in larger nuclei.

Drawbacks of the QMC methods are a restriction to local interactions, the fermion sign problem, and difficulty producing the Green's function for some types of interaction terms. These are all either addressed successfully by our implementation of the methods or avoided by our choice of interactions. A more immediate limitation is that our basis, which spans the possible particle spins and isospins at each particle configuration, grows rapidly with A, so that limitations of both computer speed and computer memory restrict us to $A \leq 12$ at present.

In the next three sections, I sketch the interactions and computational methods that we use, as well as the principal results that have been obtained. These are reviewed in much more detail in References 1 and 2.

2. Interactions

We use the Argonne v_{18} nucleon-nucleon potential[3] (AV18). It is a "realistic" potential in the sense that it fits the Nijmegen 1993 phase shift analysis[4] of nucleon-nucleon (NN) scattering up to the pion threshold with reduced χ^2 of 1.09. AV18 is a local interaction consisting of the sum of eighteen operator functions of the particle pair separation, the operators being central, spin-spin, tensor, spin-orbit, and so on. Its more important properties are a strong repulsive core, a strong tensor interaction, and a long-range form that matches pion exchange. AV18 includes many details of the electromagnetic interaction (e.g., magnetic-moment and vacuum polarization terms) as well as charge symmetry breaking and charge dependence.

In systems of $A \geq 3$, the three-nucleon interaction (TNI) provides a large fraction of the binding energies and spin-orbit splittings. Our descriptions of this interaction are based on understanding its source as the omission of explicit pion and delta-resonance degrees of freedom in the wave function, but it is also tangled inextricably with the off-shell behavior of the NN interaction. The TNI is more difficult to constrain than the pair interaction because there is no three-body analogue of the NN scattering data that are used to isolate the pair interaction.

The current generation of TNI is the Illinois family of interactions[5]. Of these, we believe Illinois-2 (IL2) to provide the best overall description of the light nuclei and nuclear matter. This interaction consists of four terms that depend on the relative coordinates, spins, and isospins of the three nucleons. The spatial dependence and operator structure of three of these terms derive from two-pion exchange and three-pion ring diagrams. The fourth term is a more phenomenological short-range repulsion; the strengths of these terms were fixed to provide the best simultaneous fit to the energies of seventeen bound and narrow levels in $A \leq 8$ nuclei. After computing energies of 38 bound and narrow states in nuclei with $A \leq 10$, we have found that the root-mean-square (RMS) deviation of these energies from the experimental energies of the states is about 750 keV[6,7].

3. Computational methods

The variational Monte Carlo (VMC) method is based on a sophisticated ansatz for the wave function, consisting of a simple shell-model-like structure modified by operator functions of the pair separations[1]:

$$\begin{aligned}\Psi_T &= [\text{3-body operator functions}] \times [\text{2-body operator functions}] \\ &\quad \times [\text{2-body scalars}] \times [\text{shell-model-like orbital/spin/isospin structure}] \,.\end{aligned} \tag{1}$$

The two-body correlations solve sets of differential equations that incorporate the potential, while the three-body correlations are based on the TNI.

Each of the functions in square brackets contains adjustable parameters. We evaluate the energy expectation value of Ψ_T by Monte Carlo integration over the $3A$ particle coordinates, producing a variational upper bound on the ground state energy for the given angular momentum, parity, and isospin (which are built into Ψ_T). We then adjust the parameters of Ψ_T by hand to minimize its energy, and finally diagonalize in a small shell-model basis. This procedure works quite well in the s shell but misses the true energies by about 1 MeV per p-shell particle.

Although the VMC wave functions have limited accuracy, they represent the true wave functions quite well for some applications like the computation of spectroscopic factors. To obtain essentially the true ground state for given quantum numbers, we use Ψ_T from VMC as an initial guess in a second method, called Green's function Monte Carlo (GFMC). GFMC is an operator method that projects the true ground state out of the VMC wave function, evolving the wave function in imaginary time τ to find[1]

$$\Psi(\tau) = \exp\left[-\left(H - \tilde{E}\right)\tau\right]\Psi_T, \tag{2}$$

where $\tau = 0$ at the start of the calculation, H is the Hamiltonian, and $\tilde{E}$ is a guess at the ground-state energy. As $\tau \to \infty$, high-energy excitations are filtered out so that $\Psi(\tau)$ approaches the ground state.

GFMC starts with many ($\sim 10^5$ at $A = 5$) randomly-chosen samples of Ψ_T, called "walkers," and propagates them to higher τ in steps of length $\Delta\tau$. At each step, a new set of particle coordinates is chosen randomly (from a known distribution) for each walker, and a Green's function derived from the potential is used to propagate the walker to its new location at the new τ. Walkers are replicated or eliminated in a Monte Carlo fashion based on their amplitudes, and their walks through configuration space are restricted by a path constraint that mitigates the fermion sign problem. The operator $\exp\left[-\left(H - \tilde{E}\right)\tau\right]$ is built up from $\tau/\Delta\tau$ successive steps, and walkers at time τ are effectively samples of $\Psi(\tau)$. Our Ψ_T differs from the ground state mostly by a very small amount of contamination from very high-energy excitations; these are filtered out very rapidly (see Eq. 2), typically before $\tau = 0.1$ MeV^{-1}. Because GFMC projects out the lowest energy level present in Ψ_T, we can also compute second, third, or higher states of given quantum numbers when Ψ_T has been constructed properly[7].

Because GFMC only produces random samples of $\Psi(\tau)$ at each τ, obtaining matrix elements requires some work. We approximate the expectation value of an operator $\mathcal{O}$ by $\langle\Psi_T|\mathcal{O}|\Psi(\tau)\rangle$ and correct it perturbatively to obtain $\langle\Psi(\tau)|\mathcal{O}|\Psi(\tau)\rangle$. After the energy has "converged," we continue propagating to higher τ and average over $\langle\mathcal{O}\rangle$ at many τ values to find statistical averages and variances from the Monte Carlo sampling. The energy should not need correction, because H commutes with the operator in Eq. 2. However, we do not know how to write the contribution to that operator from terms in H containing the square of the orbital angular momentum, so we omit such terms from the propagation and compute their contributions to the energy by a small perturbative correction.

We have computed the energies of many bound and narrow states using these methods[1,7], with results shown in Figure 1. Although AV18 was fitted only to two-body scattering, and IL2 was fitted to only seventeen states in $A \leq 8$, the results reproduce the energy spectra out to $A = 12$ in many respects. The absolute binding energies are well reproduced only when the TNI is included. Spacings and orderings of states are also well-reproduced, with the TNI playing an important role in spacing across the p-shell and also affecting the ordering at $A = 10$.

4. Beyond energy levels

Our work on nuclear QMC up to now has emphasized energy spectra, because these are the easiest observables to compute. There are many reasons that energy

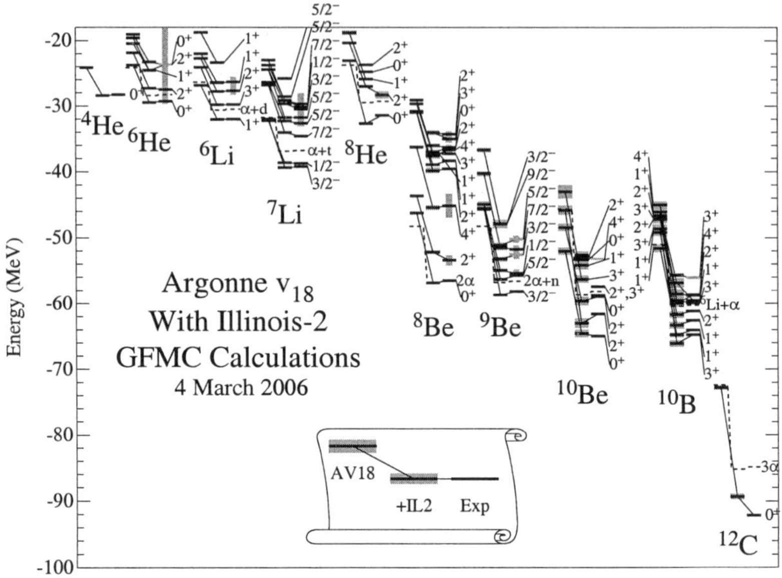

Figure 1. Calculated and measured energy levels. For each nucleus there are three energy-level diagrams. On the left is the result computed with only the two-body interaction, in the middle with the complete AV18+IL2 interaction, and on the right the measured energy. The ^{12}C result is preliminary.

spectra will be less emphasized in the coming years. One is the computational limit on A, which we are approaching with energy levels but not with other properties. In addition, there are many static, reaction, and decay properties of nuclei that can provide experimental tests of our interactions. We should also be able to provide useful information on reaction cross sections for astrophysics that cannot be measured easily.

In the past, we have examined weak decay rates[8], proton knockout from ^{7}Li by electrons[9], overlap functions and spectroscopic factors for interpretation of the stripping data[10,11], and radiative α captures[12,13], all using VMC wave functions. We have also examined nuclear charge radii using the GFMC method. In all but the most weakly-bound, diffuse nuclei, the charge radii have been reliable and have not changed since the initial GFMC calculations with the up-to-date interaction[5,6].

Charge radii of weakly-bound nuclei are currently the focus of intensive effort using GFMC. An initial round of ^{A}He RMS radii[5] that agreed well with a subsequent ^{6}He charge radius measurement[14] unfortunately suffered from numerical problems related to the diffuseness of the neutron halo: the Green's function had not been tabulated on a large enough grid. The GFMC code has been improved to avoid this and related problems, and a result good to one percent and still in fair agreement with experiment was found. However, this is still work in progress.

At about the same time, we found a way to improve the Ψ_T of ^{8}He that produced a 0.5 MeV change in our best value for the ^{8}He binding energy. Because the long-range tails of the wave functions, and thus their RMS radii, depend strongly on the neutron separation energies, this result makes AV18+IL2 a bad choice for computing the charge radius of ^{8}He. We have tentatively produced a new variation on Illinois-2, known as Illinois-6, which better reproduces the nucleon separation energies of He and Li isotopes and still fits other binding energies, with only changes of less than 10% to the strength parameters of the TNI. We hope to have firm predictions of charge radii before efforts to measure the charge radius of ^{8}He are complete.

Some years ago, we computed cross sections for three radiative α-capture reactions[12,13] of interest for astrophysical problems. There were several technical challenges to overcome, but we concentrated on learning to compute the matrix elements and to impose accurate long-range tails on the bound states. We deferred for later the problems of computing off-diagonal matrix elements from GFMC wave function samples and solving for scattering wave functions *ab initio*.

We used VMC wave functions for the bound states. Scattering wave functions, needed for the initial state (*e.g.* scattering of an α particle and deuteron in the reaction $d(\alpha,\gamma)^6$Li), were written as antisymmetrized products of the form $\phi_1\phi_2\psi_{12}(\vec{r}_{12})$. The internal wave functions ϕ_i of the scattering nuclei were found by VMC, while the correlation $\psi_{12}(\vec{r}_{12})$ corresponding to their relative motion was not derived from the underlying nucleon interactions. Instead, it was the solution of a one-body Schrödinger equation for a potential fitted to match laboratory elastic-scattering data. There are several such potentials in the literature for the reactions we examined, so we estimated the error introduced by this approach by examining the variation of the cross section from one scattering potential to another.

Results for two of the three reactions examined are shown in Fig. 2. These are in excellent agreement with the data, considering the precision of both the data and the calculations. The energy dependence is correct because we built the correct shapes into the tails of the final-state VMC wave functions. Of more interest are the absolute normalizations of the curves. They test how well VMC provides the normalizations of the wave function tails. In the case of ^{3}H$(\alpha,\gamma)^7$Li, our normalization is a good match to the best measurement[15] (the upward-pointing triangles). For ^{3}He$(\alpha,\gamma)^7$Be, our normalization is in reasonable agreement with the (mildy discordant) data, but only passes through the one or two lowest data sets. A more detailed summary of this work and related work on electroweak

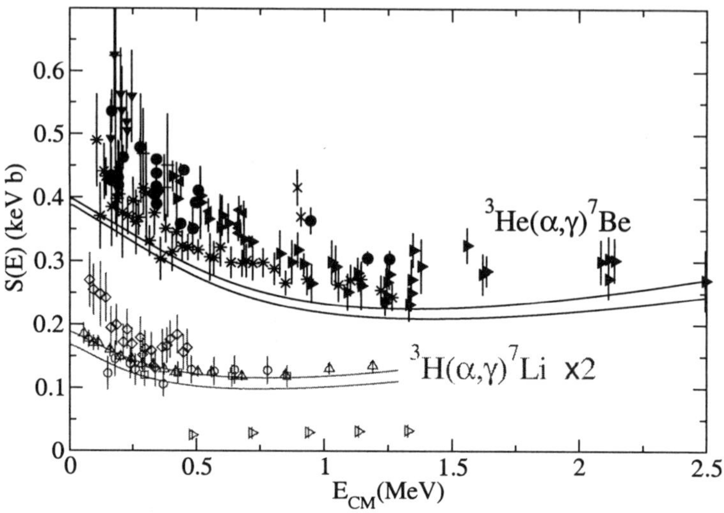

Figure 2. Astrophysical S-factors for the $^3\mathrm{He}(\alpha,\gamma)^7\mathrm{Be}$ (upper portion of the figure with darker symbols) and $^3\mathrm{H}(\alpha,\gamma)^7\mathrm{Li}$ (lower portion with lighter symbols) reactions. The solid curves show the results of VMC calculations; the space between them indicates the range allowed by our treatment of scattering. The symbols show experimental results. From Ref. 2.

processes may be found in Ref. 2.

The precision of the capture calculations is not such that disagreement with measured cross sections would indicate either shortcomings of the nuclear potential or systematic problems in the data. However, this is the goal. To reach it, we must learn to compute scattering wave functions (in both VMC and GFMC) and to compute off-diagonal matrix elements using the sampled wave functions of GFMC. In the next section, I describe a recently completed learning problem that takes us in that direction.

5. Learning to scatter: ^{5}He

We solve scattering problems via the time-independent Schrödinger equation, with the system confined to a spherical box. Specifying the logarithmic derivative of the wave function normal to the box surface gives the unbound-state problem discrete energy levels. In the language of Wigner **R**-matrix theory[16], we compute the wave function in the "internal region" where strong interactions are non-zero.

The box size is specified for our specific ^{4}He+n learning problem by the alpha-neutron separation distance, $r_{\alpha n} = R$. The parameter R is arbitrary, except that all interactions but Coulomb (when applicable) have to be zero outside the boundary. The parameter B in the condition $B\Psi = \hat{\mathbf{n}} \cdot \nabla_{\alpha n}\Psi$ at the boundary can be chosen at will. Having chosen R and B, we compute the ground-state energy E of the system (and possibly higher states in the future, following the same procedure as for bound excited states). Given the energy and the values of B and R, the wave function can be matched across the boundary onto a sum of regular and irregular functions to find the phase shift δ at the computed energy. By successive calculations with varying B, we map out $E(B)$ or equivalently $\delta(E)$. Past QMC studies of scattering in nuclear[17,18,19], condensed-matter[20] and atomic[21,22] systems obtained discrete states by setting the wave function to zero at the boundary. Although that condition is easy to implement, it requires larger and larger R (now the location of a node in the wave function) as the energy gets closer to threshold.

The logarithmic derivative boundary condition is easy to implement in VMC by building it directly into Eq. 1. The extension to nucleus-nucleus scattering should be almost as straightforward, being very similar to improvements made to the bound states in our work on radiative capture[12,13].

To apply the boundary condition in GFMC, we use a method of images. Ordinarily in GFMC, the wave function at a sampled configuration, $\Psi(\mathbf{R}', \tau + \Delta\tau)$, is found by starting from a previous sample $\Psi(\mathbf{R}, \tau)$, choosing a new configuration $\mathbf{R}'$, and finding $\Psi(\mathbf{R}', \tau + \Delta\tau) = G(\mathbf{R}', \mathbf{R}, \Delta\tau)\Psi(\mathbf{R}, \tau)$ from the Green's function G. To apply our boundary condition, we also find an image point $\mathbf{R}_I$ (uniquely determined by $\mathbf{R}$) outside the boundary. The wave function $\Psi_I(\mathbf{R}_I, \tau)$ at $\mathbf{R}_I$ is computable from $\Psi(\mathbf{R}, \tau)$ and B by extrapolation. The wave function sample $\Psi(\mathbf{R}', \tau + \Delta\tau)$ is then found by summing the contribution propagated from $\mathbf{R}$ with a contribution propagated from $\mathbf{R}_I$, so

$$\Psi(\mathbf{R}', \tau + \Delta\tau) = G(\mathbf{R}', \mathbf{R}, \Delta\tau)\Psi(\mathbf{R}, \tau) + G(\mathbf{R}', \mathbf{R}_I, \Delta\tau)\Psi_I(\mathbf{R}_I, \tau).$$

This method seems to work well, though substantial work has been required to arrive at the desired precision. The low-energy scattering problem places stringent demands on GFMC. GFMC only computes *absolute* energies, but what we want in scattering problems is the difference between the computed energy and the energy of separated nuclei. It has been assumed in past QMC calculations that 1% is a good precision for computing nuclear energies. However, if the goal is 100 keV precision just above the ^{4}He+n threshold in ^{5}He (to locate a resonance or fix the scattering length), precision of $(0.100 \text{ MeV})/(28 \text{ MeV}) \simeq 0.3\%$ is needed.

At this precision, many imperfections of the GFMC method demand attention. This is particularly true of states above threshold, because they are rather diffuse. These difficulties are closely related to those encountered in computing charge radii for heavier He isotopes. We have had to pay close attention to the role of the constrained-path propagation[23] and increase the number of unconstrained steps

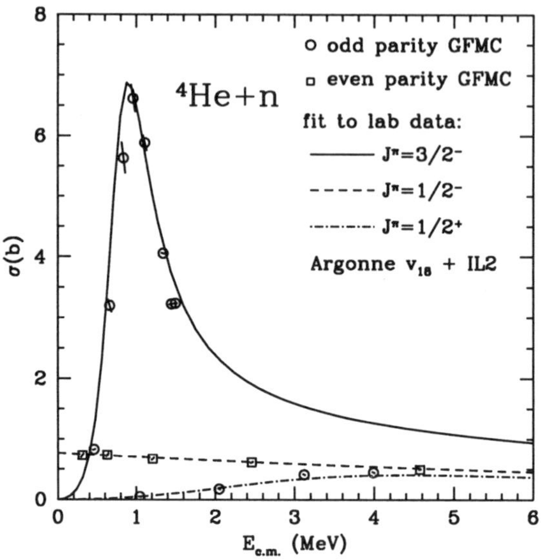

Figure 3. Scattering cross section for ^{4}He+n, decomposed into partial waves. The solid curves are the $\mathbf{R}$-matrix fit of Hale, while the points with Monte Carlo error bars are the results of GFMC calculations with the AV18+IL2 potential.

going into each energy sample. Arriving at stable answers that do not drift with τ or with minor changes in Ψ_T has also required adjusting the large-$r_{\alpha n}$ tail of Ψ_T iteratively to make it as similar as possible to the GFMC solution.

The main constraint on the boundary location R is that it must be beyond any interaction (including exchange) between the ^{4}He nucleus and the scattering neutron. This is so that scattering wave functions in the external region are sums of known functions, allowing the calculation of phase shifts. However, the highest-energy state we can reach (by dialing B) for a given box size is the first one to have a node at the boundary, and the energy of this state decreases as R increases. To cover a range of energies (and ensure that the Monte Carlo sampling is mostly in "interesting" regions of configuration space), we therefore want the smallest R consistent with no nuclear interaction at the surface. We find that our results change considerably in going from $R = 7$ fm to 8 fm, but that 9 fm and 10 fm are consistent with each other; we perform most of our calculations at 9 fm, the smallest radius consistent with correct phase shifts.

We have computed phase shifts up to a few MeV in s- and p-waves, which are the measurable contributions to scattering at those energies, using four different interactions: AV18 alone, the simplified two-body interaction AV8′, AV18 with the older Urbana IX TNI, and AV18+IL2. All of these potentials give s-wave results consistent with the data. All of the interactions give the centroid of the two p-wave resonances approximately correctly, but only IL2 produces enough splitting between the $1/2^-$ and $3/2^-$ resonances to reproduce the measured phase shifts. Fig. 3 shows the AV18+IL2 results broken down by partial wave, along with the $\mathbf{R}$-matrix fit of Hale[24], based on the available experimental data. The agreement is excellent, and derives purely from AV18+IL2 with no adjustable parameters. Moreover, while there previously was no way to compute resonance widths from GFMC, Fig. 3 shows that we can now extract them from computed phase shifts.

6. Prospects for the future

The near future holds much promise for nuclear QMC and realistic interactions, and we are moving forward with several types of calculations. We can now study broad resonances, previously inaccessible because all states had to be approximated as bound states, and in particular systems like ^{4}H, with no bound states at all. We will compute several electroweak capture cross sections, and over the long term we should be able to address nucleon-transfer reactions as well. We will soon produce off-diagonal matrix elements from GFMC, starting with beta decays.

There are (at least) two paths to relevance for nuclei with $A > 12$. One is that QMC will continue to be an important source of cross-checks for methods (e.g., no-core shell model) that are applicable across a wider range of masses. Another is that our QMC methods themselves may be adaptable to larger systems. A promising route for such a development is the auxiliary-field diffusion Monte Carlo (AFDMC) method[25,26], which replaces the explicit sums over spin-isospin states in GFMC with random sampling of auxiliary fields; much of our experience with GFMC should apply to AFDMC.

As experimental methods and facilities for studying unstable nuclei advance, there will be a wider ground for experimental tests of theory. Many quantities that we calculate now were measured decades ago, when experiment got far ahead of what could be computed *ab initio*. The coming years promise tighter coupling of *ab initio* theory with experiment, and QMC methods are particularly well-suited to the weakly-bound or unbound nuclei that will dominate the efforts of our experimental colleagues over the next several years.

Acknowledgments

The work described here has been carried out in collaboration with J. Carlson of Los Alamos National Laboratory and with S. C. Pieper and R. B. Wiringa of Argonne. The many-body calculations were performed on computers of the Laboratory Computing Resource Center at Argonne and of the National Energy Research Scientific Computing Center. This work is supported by the U. S. Department of Energy under contract No. W-31-109-ENG-38.

References

1. S. C. Pieper and R. B. Wiringa, *Ann. Rev. Nucl. Part. Sci.* **51**, 53 (2001).
2. L. Marcucci *et al.*, *Nucl. Phys. A* in press; doi:10.1016/j.nuclphysa.2004.09.008; nucl-th/0402078.
3. R. B. Wiringa, V. G. J. Stoks and R. Schiavilla, *Phys. Rev. C* **51**, 38 (1995).
4. V. G. J. Stoks *et al.*, *Phys. Rev. C* **48**, 792 (1993).
5. S. C. Pieper *et al.*, *Phys. Rev. C* **64**, 014001 (2001).
6. S. C. Pieper, K. Varga and R. B. Wiringa, *Phys. Rev. C* **66**, p. 044310 (2002).
7. S. C. Pieper, R. B. Wiringa and J. Carlson, *Phys. Rev. C* **70**, p. 054325 (2004).
8. R. Schiavilla and R. B. Wiringa, *Phys. Rev. C* **65**, 054302 (2002).
9. L. Lapikás, J. Wesseling and R. B. Wiringa, *Phys. Rev. Lett.* **82**, 4404 (1999).
10. A. H. Wuosmaa *et al.*, *Phys. Rev. C* **72**, 061301(R) (2005).
11. A. H. Wuosmaa *et al.*, *Phys. Rev. Lett.* **94**, 082502 (2005).
12. K. M. Nollett, R. B. Wiringa and R. Schiavilla, *Phys. Rev. C* **63**, p. 024003 (2001).
13. K. M. Nollett, *Phys. Rev. C* **63**, 054002 (2001).
14. L.-B. Wang *et al.*, *Phys. Rev. Lett.* **93**, 142501 (2004).
15. C. R. Brune, R. W. Kavanagh and C. Rolfs, *Phys. Rev. C* **50**, 2205 (1994).
16. A. M. Lane and R. G. Thomas, *Rev. Mod. Phys.* **30**, 257 (1958).
17. J. Carlson, V. R. Pandharipande and R. B. Wiringa, *Nuc. Phys. A* **424**, 47 (1984).
18. J. Carlson, K. E. Schmidt and M. H. Kalos, *Phys. Rev. C* **36**, 27 (1987).
19. Y. Alhassid and S. E. Koonin, *Ann. Physics* **155**, 108 (1984).
20. J. Shumway and D. M. Ceperley, *Phys. Rev. B* **63**, 165209 (2001).
21. S. Chiesa, M. Mella and G. Morosi, *Phys. Rev. A* **66**, 042502 (2002).
22. S. Chiesa, M. Mella and G. Morosi, *Phys. Rev. A* **69**, 022701 (2004).
23. R. B. Wiringa *et al.*, *Phys. Rev. C* **62**, 014001 (2000).
24. G. M. Hale, in *Proc. Int. Symp. on Nuclear Data Evaluation Methodology, Brookhaven National Laboratory, 12–16 October 1992*, ed. C. L. Dunford (World Scientific.
25. A. Sarsa *et al.*, *Phys. Rev. C* **68**, 024308 (2003).
26. S. Gandolfi *et al.*, *Phys. Rev. C* **73**, 044304 (2006).

HARMONIC-OSCILLATOR-BASED EFFECTIVE THEORY

W. C. HAXTON

Inst. for Nuclear Theory and Dept. of Physics, University of Washington
Seattle, WA 98195, USA
E-mail: haxton@phys.washington.edu

I describe harmonic-oscillator-based effective theory (HOBET) and explore the extent to which the effects of excluded higher-energy oscillator shells can be represented by a contact-gradient expansion in next-to-next-to-leading order (NNLO). I find the expansion can be very successful provided the energy dependence of the effective interaction, connected with missing long-wavelength physics associated with low-energy breakup channels, is taken into account. I discuss a modification that removes operator mixing from HOBET, simplifying the task of determining the parameters of an NNLO interaction.

1. Introduction

Often the problem of calculating long-wavelength nuclear observables – binding energies, radii, or responses to low-momentum probes – is formulated in terms of pointlike, nonrelativistic nucleons interacting through a potential. To solve this problem theorists have developed both nuclear models, which are not systematically improvable, and exact numerical techniques, such as fermion Monte Carlo. Because the nuclear many-body problem is so difficult – one must simultaneously deal with anomalously large NN scattering lengths and a potential that has a short-range, strongly repulsive core – exact approaches are numerically challenging, so far limited to the lighter nuclei within the 1s and lower 1p shells. The Argonne theory group has been one of the main developers of such exact methods [1].

However, effective theory (ET) offers an alternative, a method that limits the numerical difficulty of a calculation by restricting it to a finite Hilbert space (the P- or "included"-space), while correcting the bare Hamiltonian H (and other operators) for the effects of the Q- or "excluded"-space. Calculations using the effective Hamiltonian H^{eff} within P reproduce the results using H within $P+Q$, over the domain of overlap. That is, the effects of Q on P-space calculations are absorbed into $P(H^{eff} - H)P$. There may exist some systematic expansion – perhaps having to do with the shorter range of interactions in Q – that simplifies the determination $P(H^{eff} - H)P$ [2,3].

One interesting challenge for ET is the case of a P-space basis of harmonic oscillator (HO) Slater determinants. This is a special basis for nuclear physics because of center-of-mass separability: if all Slater determinants containing up to n oscillator quanta are retained, H^{eff} will be translationally invariant (assuming H is). Such bases are also important because of powerful shell-model (SM) techniques that have been developed for interative diagonalization and for evaluating inclusive responses. The larger P can be made, the smaller the effects of $H^{eff} - H$.

There are two common approaches to the ET problem. One is the determination of $P(H^{eff} - H)P$ from a given H known throughout $P+Q$, a problem that appears naively to be no less difficult than the original $P+Q$ diagonalization of H. However, this may not be the case if H is somehow simpler when acting in Q. For example, if Q contains primarily high-momentum (short-distance) interactions, then $H^{eff} - H$ might have a cluster expansion: it becomes increasingly unlikely to have m nucleons in close proximity, as m increases (e.g., a maximum of four nucleons can be in a relative s-state). Thus one could approximate the full scattering series in Q by successive two-body, three-body, etc., terms, with the expectation that this series will converge quickly with increasing nucleon number. This would explain why simple two-nucleon ladder sums – g-matrices – have been somewhat successful as effective interactions [4] (however, see Ref. 5, 6).

The second approach is that usually taken in effective field theories [3], determining H^{eff} phenomenologically. This is the "eliminate the middleman" approach: H itself is an effective interaction, parameterized in order to reproduce NN scattering and other data up to some energy. So why go to the extra work of this intermediate stage between QCD and SM-like spaces? This alternative approach begins with PHP, the long-range NN interaction that is dominated by pion exchange and constrained by chiral symmetry. The effects of the omitted Q-space, $P(H^{eff} - H)P$, might be expressed in some systematic expansion, with the coefficients of that expansion directly determined from data, rather than from any knowledge of H acting outside of P.

While we explored this second approach some years ago, some subtle issues arose, connected with properties of HO bases. This convinced us that the first step in our program should be solving and thoroughly understanding the effective interactions problem via the first approach, so that we would have answers in hand to test the success of more phenomenological approaches. Thus we proceeded to follow the first approach using a realistic NN potential, $av18$ [7], generating H^{eff} numerically for the two- and three-body problems in a variety of HO SM spaces. Here I will use these results to show that a properly defined short-range interaction provides an excellent representation of the effective interaction. This is an encouraging result, one that suggests a purely phenomenological treatment of the effective interaction might succeed.

The key observation is that HOBET is an expansion around momenta $k \sim 1/b$, and thus differs from EFT approaches that expand around $k \sim 0$. Consequently a HOBET P-space lacks both high-momentum components important to short-range NN interactions and long-wavelength components important to minimizing the kinetic energy. While our group has previously discussed some of the consequences of the combined infrared/ultraviolet problem in HOBET, here I identify another: a sharp energy dependence in H^{eff} that must be addressed before any simple representation of $H^{eff} - H$ is possible. This, combined with a trick to remove operator mixing, leads to a simple and successful short-range expansion for $H^{eff} - H$. I conclude by noting how these results may set the stage for a successful determination of the HOBET H^{eff} directly from data.

2. Review of the Bloch-Horowitz Equation

The basis for the approach described here is the Bloch-Horowitz (BH) equation, which generates a Hermitian, energy-dependent effective Hamiltonian, H^{eff}, which operates in a finite Hilbert space from which high-energy HO Slater determinants are omitted:

$$H^{eff} = H + H \frac{1}{E - QH} QH$$

$$H^{eff}|\Psi_P\rangle = E|\Psi_P\rangle \qquad |\Psi_P\rangle = (1-Q)|\Psi\rangle. \tag{1}$$

Here H is the bare Hamiltonian and E and Ψ are the exact eigenvalue and wave function (that is, the results of a full solution of the Schroedinger equation for H in $P+Q$). The BH equation must be solved self-consistently, as H^{eff} depends on E. If this is done, the model-space calculation reproduces the exact E, and the model-space wave function Ψ_P is simply the restriction of Ψ to P. If one takes for P a complete set of HO Slater determinants with HO energy $\leq \Lambda_P \hbar\omega$, H^{eff} will be translational invariant. P is then defined by two parameters, Λ_P and the HO size parameter b.

The BH equation was solved numerically for the $av18$ potential using two numerical techniques. In work carried out in collaboration with C.-L. Song [8], calculations were done for the deuteron and ^{3}He/^{3}H by directly summing the effects of $av18$ in the Q-space. Because this potential has a rather hard core, sums to 140 $\hbar\omega$ were required to achieve $\sim$ 1 keV accuracy in the deuteron binding energy, and 70 $\hbar\omega$ to achieve $\sim$ 10 keV accuracies for ^{3}He/^{3}H. In work carried out with T. Luu [9], such cutoffs were removed by doing momentum-space integrations over all possible excitations.

The results are helpful not only to the goals discussed previously, but also in illustrating general properties of H^{eff} that may not be widely appreciated. For example, Table 1 gives the evolution of the P-space ^{3}He $av18$ wave function as a function of increasing Λ_P for fixed b. Ψ_P evolves simply, with each increment of Λ_P adding new components to the wave function, while leaving previous components unchanged. One sees that the probability of residing in the model space grows slowly from its $0\hbar\omega$ value (31%) toward unity.

Table 1. Evolution of the ^{3}He $av18$ HO wave function Ψ_P with Λ_P

state	amplitude					
	$0\hbar\omega$	$2\hbar\omega$	$4\hbar\omega$	$6\hbar\omega$	$8\ \hbar\omega$	exact
	(31.1%)	(57.4%)	(70.0%)	(79.8%)	(85.5%)	(100%)
$\mid 0,1\rangle$	0.5579	0.5579	0.5579	0.5579	0.5579	0.5579
$\mid 2,1\rangle$	0.0000	0.0463	0.0461	0.0462	0.0462	0.0463
$\mid 2,2\rangle$	0.0000	-0.4825	-0.4824	-0.4824	-0.4824	-0.4826
$\mid 2,3\rangle$	0.0000	0.0073	0.0073	0.0073	0.0073	0.0073
$\mid 4,1\rangle$	0.0000	0.0000	-0.0204	-0.0204	-0.0204	-0.0205
$\mid 4,2\rangle$	0.0000	0.0000	0.1127	0.1127	0.1127	0.1129
$\mid 4,3\rangle$	0.0000	0.0000	-0.0419	-0.0420	-0.0421	-0.0423

Effective operators are defined by

$$\hat{O}^{eff} = (1 + HQ\frac{1}{E_f - HQ})\hat{O}(1 + \frac{1}{E_i - QH}QH) \tag{2}$$

and must be evaluated between wave functions Ψ_P having the nontrivial normalization illustrated in Table 1 and determined by

$$1 = \langle\Psi|\Psi\rangle = \langle\Psi_P|\hat{1}^{eff}|\Psi_P\rangle. \tag{3}$$

The importance of this is illustrated in Fig. 1, where the elastic magnetic responses for deuterium and ^{3}He are first evaluate with exact wave functions Ψ_P but bare operators, then re-evaluated with the appropriate effective operators. Bare operators prove a disaster even at intermediate momentum transfers of 2-3 f^{-1}. By using the effective operator and effective wave function appropriate to Λ_P, the correct result – the form factor is independent of the choice of Λ_P (or b) – is obtained, as it must in any correct application of effective theory.

The choice of a HO basis excludes not only high-momentum components of wave functions connected with the hard core, but also low-momentum components connected with the proper asymptotic fall-off of the tail of the wave func-

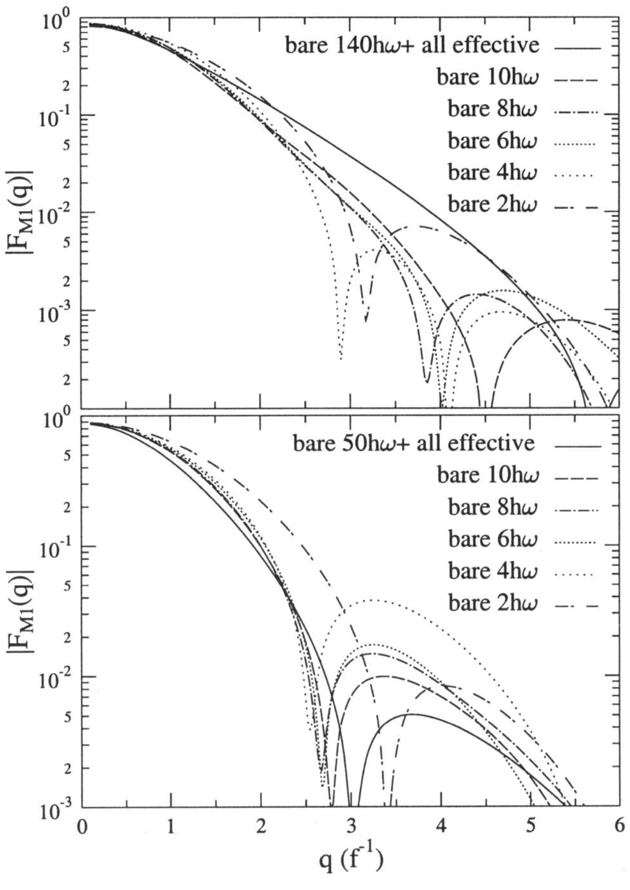

Figure 1. Deuteron and ^{3}He elastic magnetic form factors evaluated for various P-spaces with bare operators (various dashed and dotted lines) and with the appropriate effective operators (all results converge to the solid lines).

tions. This combined ultraviolet/infrared problem was first explored by us in connection with the nonperturbative behavior of H^{eff}: the need to simultaneously correct for the missing long- and short-distance behavior of Ψ_P is the reason one cannot define a simple P that makes evaluation of H^{eff} converge rapidly. We also found a solution to this problem, a rewriting of the BH equation in which the relative kinetic energy operator is summed to all orders. This summation can be viewed as a transformation of a subset of the Slater determinants in P to incorporate the correct asymptotic falloff. This soft physics, obtained from an infinite sum of high-energy HO states in Q, is the key to making H^{eff} perturbative.

It turns out this physics is also central to the issue under discussion here, the existence of a simple representation for $H^{eff} = H + H\frac{1}{E-QH}QH$, where $H = T + V$. The reorganized BH H^{eff} is the sum of the three left-hand-side (LHS) terms in Eqs. (4)

$$\langle\alpha|T + TQ\frac{1}{E-QT}QT|\beta\rangle \underset{nonedge}{\longrightarrow} \langle\alpha|T|\beta\rangle$$
$$\langle\alpha|\frac{E}{E-TQ}V\frac{E}{E-QT}|\beta\rangle \underset{nonedge}{\longrightarrow} \langle\alpha|V|\beta\rangle$$
$$\langle\alpha|\frac{E}{E-TQ}V\frac{1}{E-QH}QV\frac{E}{E-QT}|\beta\rangle \underset{nonedge}{\longrightarrow} \langle\alpha|V\frac{1}{E-QH}QV|\beta\rangle \quad (4)$$

The first LHS term is the effective interaction for T, the relative kinetic energy. As QT acts as a ladder operator in the HO, $E/E-QT$ is the identity except when it operates on an $|\alpha\rangle$ with energy $\Lambda_P\hbar\omega$ or $(\Lambda_P-1)\hbar\omega$. We will call these Slater determinants the edge states. For nonedge states, this new expression and the BH form given in Eq. (1) both reduce to the expressions on the right of Eqs. (4).

Noting that the first LHS term in Eqs. (4) can be rewritten as

$$\langle\alpha|\frac{E}{E-TQ}(T - \frac{TQT}{E})\frac{E}{E-QT}|\beta\rangle, \quad (5)$$

we see that the QT summation can be viewed as a transformation to a new basis for P, $E/(E-QT)|\alpha\rangle$, that is orthogonal but not orthonormal. This edge-state basis builds in the proper asymptotic behavior governed by QT (free propagation) and the binding energy E. The transformation preserves translational invariance, as T is the relative kinetic energy operator. Viewed in the transformed basis, the appropriate effective interaction in given by the LHS terms in the square brackets in Eqs. (6) below.

Alternatively, the results can be viewed as two equivalent expressions for the effective interaction between HO states, but with a different division between "bare" and "rescattering" contributions

$$\text{bare}: \frac{E}{E-TQ}\left[H - \frac{TQT}{E}\right]\frac{E}{E-QT} \Leftrightarrow H$$
$$\text{rescattering}: \frac{E}{E-TQ}\left[V\frac{1}{E-QH}QV\right]\frac{E}{E-QT} \Leftrightarrow H\frac{1}{E-QH}QH \quad (6)$$

It is this new division that is critical. The expressions are identical for nonedge states. But for edge states, only the expression on the left isolates a quantity, VGV, that is short-range and nonperturbative. We will see that this is the term that can be represented by a simple, systematic expansion.

Figure 2 shows the extended tail that is induced by $E/E-QT$ acting on a HO state. Figure 3 is included to emphasize that there are important numerical differences between the two expressions in Eqs. (6). It compares calculations done for the deuteron using the two "bare" interactions: thus in both cases V enters only linearly between low-momentum states, and all multiple scattering of V in Q in ignored. Figure 3 gives the resulting deuteron binding energy as a function of b, for several values of Λ_P. For the standard form of the BH equation, a small model space overestimates the kinetic energy (too confining) and overestimates short-range contributions to V (too little freedom to create the needed wave-function "hole"). Making b larger to lower the kinetic energy exacerbates the short-range problem, and conversely. Thus the best b is a poor compromise that, even in a 10 $\hbar\omega$ bare calculation, fails to bind the deuteron. But the new bare H on the LHS of Eqs. (6) sums QT to give the correct wave-function behavior at large r, independent of b. Then, for the choice $b \sim$ 0.4-0.5 f, the short-range physics can be absorbed directly into the P space. The result is excellent 0th-order ground-state energy, with the residual effects of multiple scattering through QV being very small and perturbative [9].

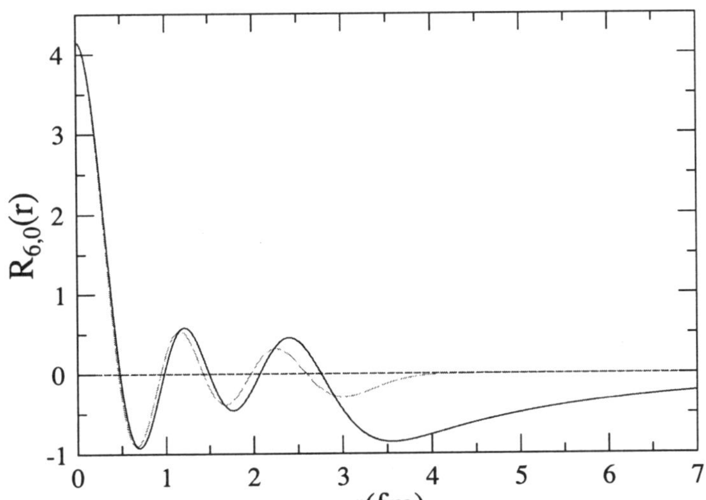

Figure 2. A comparison of the $|nl\rangle$ and extended $(E/E - QT)|nl\rangle$ radial wave functions, for the edge state $(n, l) = (6, 0)$ in a $\Lambda_P = 10$ deuteron calculation. Note that the normalization of the extended state has been adjusted to match that of $|nl\rangle$ at r=0, in order to show that the shapes differ only at large r. Thus the depletion of the extended state at small r is not apparent in this figure.

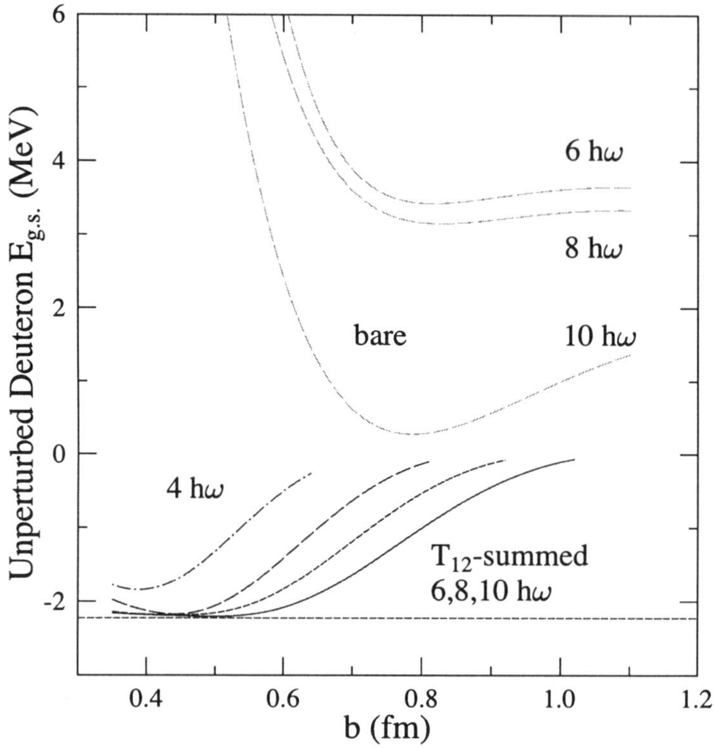

Figure 3. Deuteron ground-state convergence in small P-spaces, omitting all effects due to the multiple scattering of V in Q. The standard BH formulation with $P(T+V)P$ fails to bind the deuteron, even with $\Lambda_P = 10$. The reorganized BH equation, where QT has been summed to all orders but V still appears only linearly, reproduces the correct binding energy for Λ_P=6.

3. Harmonic-Oscillator-Based Effective Theory

Now I turn to the question of whether (and how) the Q-space rescattering contribution to H^{eff} might be expressed through some systematic short-range expansion. There are two steps important in applying such an expansion to HOBET. One has to do with the form of the short-range expansion. A contact-gradient (CG) expansion, constructed to include all possible LO (leading order), NLO (next-to-leading order) , NNLO (next-to-next-to-leading order), ..., interactions is commonly used,

$$a_{LO}^{ss}(\Lambda_P, b)\delta(\mathbf{r}) + a_{NLO}^{ss}(\Lambda_P, b)(\overleftarrow{\nabla}^2\delta(\mathbf{r}) + \delta(\mathbf{r})\overrightarrow{\nabla}^2) +$$
$$a_{NNLO}^{ss,22}(\Lambda_P, b)\overleftarrow{\nabla}^2\delta(\mathbf{r})\overrightarrow{\nabla}^2 + a_{NNLO}^{ss,40}(\Lambda_P, b)(\overleftarrow{\nabla}^4\delta(\mathbf{r}) + \delta(\mathbf{r})\overrightarrow{\nabla}^4). \quad (7)$$

Because HOBET is an expansion around a typical momentum scale $\sim\ 1/b$, rather than around $\vec{k}=0$, it is helpful to redefine the derivatives appearing in the CG expansion Noting

$$\overrightarrow{\nabla}^n \exp i\vec{k}\cdot\vec{r}\,|_{k=0} = 0, n = 1, 2, ..., \tag{8}$$

we demand by analogy in HOBET

$$\overrightarrow{\nabla}^n \psi_{1s}(b) = 0, n = 1, 2, ... \tag{9}$$

This can be accomplished by redefining the operators $\hat{O}$ of Eqs. (7) by

$$\hat{O} \rightarrow e^{r^2/2}\hat{O}e^{r^2/2} \tag{10}$$

The gradients in Eq. (7) then act on polynomials in r, a choice that removes all operator mixing. That is, if a_{LO} is fixed in LO to the $n=1$ to $n=1$ matrix element, where n is the nodal quantum number, it remains fixed in NLO, NNLO, etc. Furthermore, the expansion is in nodal quantum numbers, e.g.,

$$\overrightarrow{\nabla}^2 \sim (n-1) \qquad \overrightarrow{\nabla}^4 \sim (n-1)(n-2) \tag{11}$$

so that matrix elements become trivial to evaluate in any order. It can be shown that the leading order in n contribution agrees with the plane-wave result, and that operator coefficients are a generalization of standard Talmi integrals for nonlocal potentials, e.g.,

$$a_{NNLO}^{ss,22} \sim \int_0^\infty \int_0^\infty e^{-r_1^2} r_1^2 V(r_1, r_2) r_2^2 e^{-r_2^2} r_1^2 r_2^2 dr_1 dr_2 \tag{12}$$

The next step is to identify that quantity in the BH equation that should be identified with the CG expansion. This has to do with the two forms of the BH equation discussed previously. Consider the process of progressively integrating out Q in favor of the CG expansion, beginning at $\Lambda >> \Lambda_P$ and progressing to $\Lambda = \Lambda_P$. Using the projection operator

$$Q_\Lambda = \sum_{\alpha=\Lambda_P+1}^{\Lambda} |\alpha\rangle\langle\alpha| \quad Q_{\Lambda_P} \equiv 0 \tag{13}$$

we can isolate the contributions, above some scale Λ, to the two BH rescattering terms of Eqs. (6)

$$\begin{aligned}\Delta(\Lambda) &= H\frac{1}{E-QH}QH - H\frac{1}{E-Q_\Lambda H}Q_\Lambda H \\ \Delta_{QT}(\Lambda) &= \frac{E}{E-TQ}\left[V\frac{1}{E-QH}QV - V\frac{1}{E-Q_\Lambda H}Q_\Lambda V\right]\frac{E}{E-QT}.\end{aligned} \tag{14}$$

The goal of a CG expansion might be successful reproduction of the matrix elements of $\Delta(\Lambda)$ – the Q-space rescattering contributions for the standard form of the BH equation – as $\Lambda \rightarrow \Lambda_P$. This would allow us to replace all Q-space rescattering by a systematic short-range expansion, opening the door to a purely phenomenological determination of H^{eff} for the SM. The test case will be an $8\hbar\omega$ P-space calculation for the deuteron (Λ_P = 8, b=1.7 f). The running of the 15 independent $^3S_1$ matrix elements of $\Delta(\Lambda)$ are plottted in Fig. 4a. Five of these are distinguished because they involve an edge-state bra or ket (or both). The evolution of these contributions with Λ is seen to be somewhat less regular than that of nonedge-state matrix elements. The results for $\Lambda = \Lambda_P$ show that rescattering is responsible for typically 12 MeV of binding energy.

The CG fit to the results in Fig. 4a were done in LO, NLO, and NNLO as a function of Λ, using the standard form of the BH equation. The coefficients are fit to the lowest-energy matrix elements. Thus in LO $a_{LO}(\Lambda)$ is fixed by the $1s-1s$ matrix element, leaving 14 unconstrained matrix elements; the NNLO fit ($1s-1s$, $1s-2s$, $1s-3s$, and $2s-2s$) leaves 11 matrix elements unconstrained. This is easily done, because the operators do not mix; e.g., among these four, only the $1s-3s$ matrix element is influenced by $a^{ss,40}_{NNLO}$. The result is a set of coefficient that run as a function of Λ in the usual way, with a_{LO} small and dominant for large Λ, and with the NLO and NNLO terms turning on as the scale is dropped. Figs. 4b-d show the residuals – the differences between the predicted and calculated matrix elements. For non-edge-state matrix elements the scale at which typical residuals in Δ are significant, say greater than 100 keV (above $\sim 1\%$), is brought down successively, e.g., from $\sim 100\hbar\omega$, to $\sim 55\hbar\omega$ (LO), to $\sim 25\hbar\omega$ (NLO), and finally to $\sim \Lambda_P\hbar\omega$ (NNLO). But matrix elements involving edge states break this pattern: the improvement is not significant, with noticeable deviations remaining at $\sim 100\hbar\omega$ even at NNLO.

This failure could be anticipated: because QT strongly couples nearest shells across the $P-Q$ boundary, $H\frac{1}{E-QH}QH$ contains long-range physics. The candidate short-range interaction is $V\frac{1}{E-QH}QV$, not $H\frac{1}{E-QH}QH$: this is the reason QT should be first summed, putting the BH equation in a form – the LHS of Eqs. (6) – that isolates this quantity. This reorganization affects edge-state matrix elements only, those with the large residuals in Figs. 4b-d.

To use the reorganized BH equation, the QT sums appearing in Eqs. (4) must be completed. There are several procedures for doing this, but one convenient method exploits the raising/lowering properties of T. The result is a series of continued fractions $\tilde{g}_i(2E/\hbar\omega, \{\alpha_i\}, \{\beta_i\})$, where $\alpha_i = (2n+2i+l-1/2)/2$ and $\beta_i = \sqrt{(n+i)(n+i+l+1/2)}/2$. For any operator $\hat{O}$ (e.g., V, $V\frac{1}{E-QH}QV$,

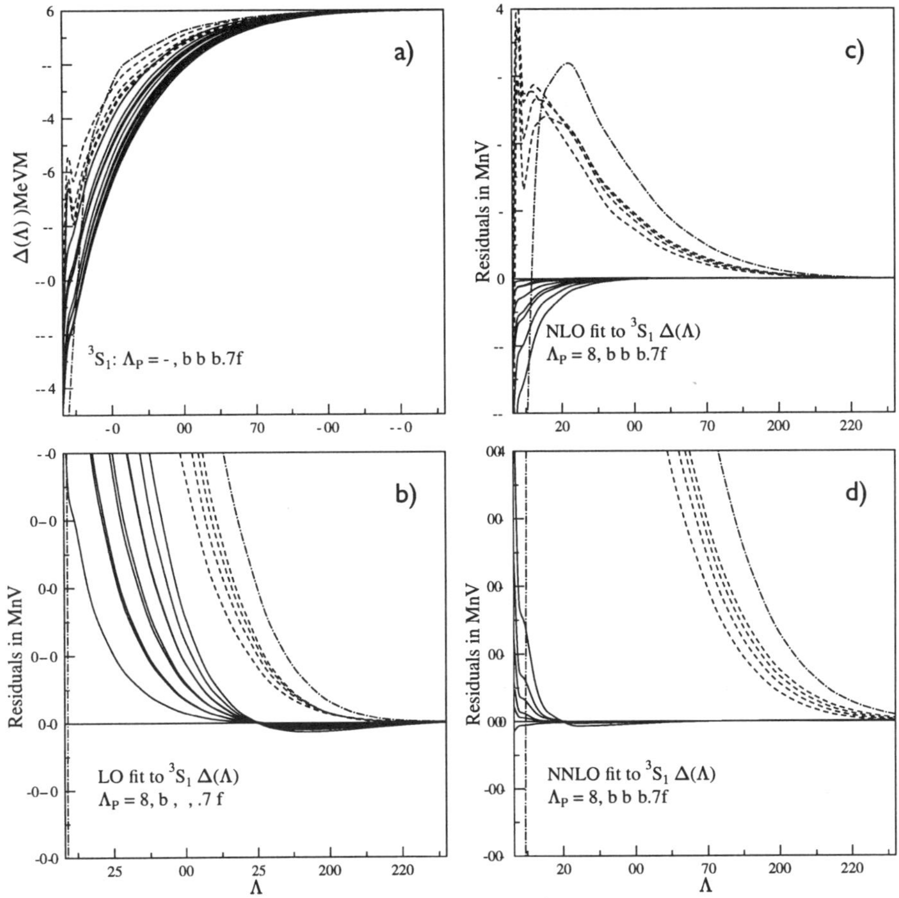

Figure 4. In a) rescattering contributions to H^{eff} due to excitations in Q above Λ are given for the standard form of the BH equation. These results are for the 15 matrix elements that arise in an $8\hbar\omega$ calculation for the deuteron, with b=1.7 f. In b)-d) the residuals of LO, NLO, and NNLO fits are shown (see text). Matrix elements with bra or ket (dashed) or both (dot-dashed) edge states are seen not to improve systematically.

etc.)

$$\langle n'l'|\frac{E}{E-TQ}O\frac{E}{E-QT}|nl\rangle = \sum_{i,j=0} \tilde{g}_j(n',l')\tilde{g}_i(n,l)\langle n'+j\ l|O|n+i\ l\rangle \quad (15)$$

It follows that the coefficients of the CG expansion for a HO basis must be redefined for edge states, with a state- and E-dependent renormalization

$$a_{LO} \to a'_{LO}(E/\hbar\omega; n', l', n, l) = a_{LO} \sum_{i,j=0} \tilde{g}_j(n', l')\tilde{g}_i(n, l)$$
$$\times \left[\frac{\Gamma(n'+j+1/2)\Gamma(n+i+1/2)}{\Gamma(n'+1/2)\Gamma(n+1/2)}\right]^{1/2} \left[\frac{(n'-1)!(n-1)!}{(n'+j-1)!(n+i-1)!}\right]^{1/2} \quad (16)$$

This renormalization, which introduces no new parameters, can be evaluated in a similar way for heavier systems: T remains a raising operator.

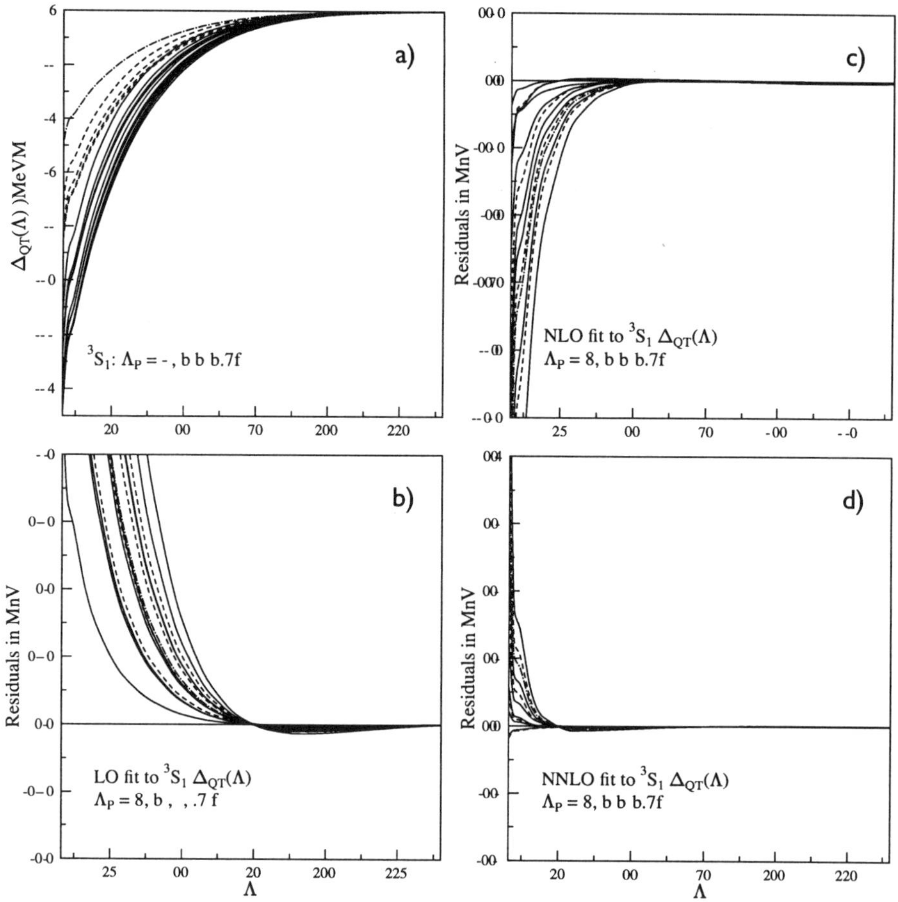

Figure 5. As in Fig. 4, but with the edge states treated according to the QT-summed reorganization of the BH equation, as described in the text.

Figs. 5 shows the results: the difficulties encountered for $\Delta(\Lambda)$ do not arise for $\Delta_{QT}(\Lambda)$. The edge-state matrix elements are now well behaved, and the improvement from LO to NLO to NNLO is systematic in all cases. When $\Lambda \to \Lambda_P$, the CG potential continues to reproduce H^{eff} for the $av18$ potential remarkably well, with an the average error in $^3S_1$ NNLO matrix elements of about 100 keV (or 1% accuracy). Other channels we explored behaved even better: the average error for the 15 $^1S_0$ matrix elements is about 10 keV (or 0.1% accuracy). Because all matrix elements of H^{eff} are reproduced well, the CG potential preserves spectral properties, not simply properties of the lowest energy states within P. The NNLO calculation in the $^3S_1$-$^3D_1$ channel yields a deuteron ground-state energy of -2.21 MeV.

Several points can be made:

• The net effect of the QT summation is to weaken the CG potential for HO edge states: the resulting, more extended state has a reduced probability at small r. Consequently the effects of QV are weaker than in states immune from the effects of QT.

• The very strong QT coupling of the P and Q spaces is clearly problematic for an ET: small changes in energy denominators alter the induced interactions. Thus it is quite reasonable that removal of this coupling leads to a strong energy dependence in the effective interaction between HO states. I believe that proper treatment of this energy dependence will be crucial to a correct description of the bound-state spectrum in the HO SM.

• This process can also be viewed as a transformation to a new, orthogonal (but not normalized) basis for P in which $|\alpha\rangle \to \frac{E}{E-QT}|\alpha\rangle$. This yields basis states with the proper asymptotic behavior for each channel. A CG expansion with fixed coefficients can be used between these states, following the reorganized BH equation of Eqs. (4).

• While our calculations have been limited to the deuteron, this same phenomena must arise in heavier systems treated in HO bases – QT remains the ladder operator. As the issue is extended states that minimize the kinetic energy, it is clear that the relevant parameter must be the Jacobi coordinate associated with the lowest breakup channel. This could be an issue for treatments of H^{eff} based on the Lee-Suzuki transformation, which transforms the interaction into an energy-independent one . In approaches like the no-core shell model [10], the Lee-Suzuki transformation is generally not evaluated exactly, but instead only at the two-body level. If such an approach were applied, for example, to ^{6}Li, a system weakly bound (1.475 MeV) to breaking up as α+d, it is not obvious that a two-body Lee-Suzuki transformation would treat the relevant Jacobi coordinate responsible for the dominant energy dependence. This should be explored.

• I believe the conclusions about the CG expansion will apply to other effective interactions. For example, V-low-k [11], a soft potential obtained by integrating out high-momentum states, is derived in a plane-wave basis, where T is diagonal. Thus it should be analogous to our CG interaction, requiring a similar renormalization when embedded into a HO SM space. It would be interesting to test this conclusion.

4. Summary

These results show that the effective interaction in the HO SM must have a very sharp dependence on the binding energy, defined as the energy of the bound state relative to the first open channel. This is typically 0 to 10 MeV for the bound states of most nuclei (and 2.22 MeV for the deuteron ground state explored here). Once this energy dependence is identified, the set of effective interaction matrix elements can be represented quite well by a CG expansion, and the results for successive LO, NLO, and NNLO calculations improve systematically.

This result suggests that the explicit energy dependence of the BH equation is almost entirely due to QT – though this inference, based on the behavior of matrix elements between states with different number of HO quanta, must be tested in a case where multiple bound states exist.

We also presented a simple redefinition of the gradients associated with CG expansions, viewing the expansion as one around a momentum scale $\sim 1/b$. This definition removes operator mixing, making NNLO and higher-order fits very simple. The expansion then becomes one in nodal quantum numbers, with the coefficients of the expansion related to Talmi integrals, generalized for nonlocal interactions.

While our exploration here has been based on "data" obtained from an exact BH calculation of the effective interaction for the $av18$ potential, this raises the question, is such a potential necessary to the SM? That is, now that the success of an NNLO description of Q-space contributions is established, could one start with PHP and determine the coefficients for such a potential directly from data, without knowledge of matrix elements of H outside of P? I believe the answer is yes, even in cases (like the deuteron) when insufficient information is available from bound states. It turns out that the techniques described here can be extended into the continuum, so that observables like phase shifts could be combined with bound-state information to determine the coefficients of such an expansion. An effort of this sort is in progress.

I thank M. Savage for helpful discussions, and T. Luu and C.-L. Song for enjoyable collaborations. This work was supported by the U.S. Department of

Energy Division of Nuclear Physics and by DOE SciDAC grant DE-FG02-00ER-41132.

References

1. S. C. Pieper and R. B. Wiringa, *Ann.Rev.Nucl.Part.Sci.* **51**, 53 (2001).
2. S. Weinberg, *Phys. Lett.* **B251**, 288 (1990), *Nucl. Phys.* **B363**, 3 (1991), and *Phys. Lett.* **295**, 114 (1992).
3. S. R. Beane, P. F. Bedaque, W. C. Haxton, D. R. Philips, and M. J. Savage, in *At the Frontier of Particle Physics, Vol. 1* (World Scientific, Singapore, 2001), p. 133.
4. T. T. S. Kuo and G. E. Brown, *Nucl. Phys.* **A114**, 241 (1968).
5. B. R. Barrett and M. W. Kirson, *Nucl. Phys.* **A148**, 145 (1970).
6. T. H. Shucan and H. A. Wiedenmüller, *Ann. Phys. (N.Y.)* **73**, 108 (1972) and **76**, 483 (1973).
7. R. B. Wiringa, V. Stoks, and R. Schiavilla, *Phys. Rev.* **C51**, 28 (1995).
8. W. C. Haxton and C.-L. Song, *Phys. Rev. Lett.* **84**, 5484 (2000).
9. W. C. Haxton and T. C. Luu, *Nucl. Phys.* **A690**, 15 (2001) and *Phys. Rev. Lett.* **89**, 182503 (2002); T.C. Luu, S. Bogner, W. C. Haxton, and P. Navratil, *Phys. Rev.* C70, 014316 (2004).
10. P. Navratil, J. P. Vary, B. R. Barrett, *Phys. Rev. Lett.* **84**, 5728 (2000).
11. A. Schwenk, G. E. Brown, and B. Friman, *Nucl. Phys.* **A703**, 745 (2002); A. Schwenk, *J. Phys.* **G31**, S1273 (2005).

APPLICATIONS OF CONTINUUM SHELL MODEL

A. VOLYA
Department of Physics, Florida State University,
Tallahassee, FL 32306-4350, USA
E-mail: volya@phy.fsu.edu

The nuclear many-body problem at the limits of stability is considered in the framework of the Continuum Shell Model that allows a unified description of intrinsic structure and reactions. Technical details behind the method are highlighted and practical applications combining the reaction and structure pictures are presented.

1. Introduction

In this presentation we discuss specific features of the Continuum Shell Model (CoSMo), the approach based on the projection formalism [1], formulated in the classical book [2] and developed into a practical instrument in Refs. [3,4]. The whole problem of many-body physics on the verge of stability has been extensively explored in the past, especially in relation to weakly bound nuclei. Alternative formulations and their first applications can be found, for example, in Refs. [5,6,7].

The goal of this paper is to highlight complimentary views on the nuclear many-body physics from the "inside" (structure) and "outside" (reactions) perspectives. The structure view is based on the traditional shell model where the effective Hamiltonian to be diagonalized plays the central role. New contributions to the effective Hamiltonian coming from the presence of continuum bring in non-Hermiticity and energy dependence. Overcoming these complications, it is possible to calculate in the same framework the cross sections of reactions, with their energy dependence and possible resonance behavior. The complementary picture that appears from the side of nuclear reactions is important for identifying resonances and comparison with experiment. While the shell model approach to the many-body structure in discrete spectrum is firmly established the many-body reaction physics is usually left for more phenomenological tools of the reaction practitioners. The purpose of Secs. 3.1 and 3.2 in this work is to accentuate on novel methods involved in calculation of Green's functions and associated time evolution operators that stay behind CoSMo.

The example of a realistic application presented in the last section is a central point of the paper. The chain of helium isotopes is shown where a single picture combines different methods and different points of view on the same problem. The bound states of the conventional shell model below threshold are followed at higher energies by the solutions of the CoSMo effective Hamiltonian revealing resonances that coincide with the complex poles of the scattering matrix. The same resonances appear in the neutron scattering cross section plotted in the same figure. The discrepancies between the cross section peaks and resonance states emphasize subtle features of many-body dynamics in a marginally stable system.

2. Structure

Using the projection formalism one can eliminate the part of the Hilbert space related to particle(s) in continuum. This results in the effective Hamiltonian $\mathcal{H}$ that acts only in the "intrinsic" shell model space,

$$\mathcal{H}(E) = H_0 + \Delta(E) - \frac{i}{2}W(E). \tag{1}$$

Here the full Hamiltonian H_0 is restricted to intrinsic space, and is supplemented with the Hermitian term $\Delta(E)$ that describes virtual particle excitations into excluded space and the imaginary term $W(E)$ representing irreversible decays to the continuum. The new parts of the Hamiltonian (1) are found in terms of the matrix elements of the full original Hamiltonian that link the internal states $|1\rangle$ with the energy-labeled external states $|c;E\rangle$: $A_1^c(E) = \langle 1|H_0|c;E\rangle$,

$$\Delta_{12}(E) = \text{P.v.} \int dE' \sum_c \frac{A_1^c(E')A_2^{c*}(E')}{E-E'}, \quad W_{12}(E) = 2\pi \sum_{c(\text{open})} A_1^c A_2^{c*}, \tag{2}$$

Reduction of the effective space does not go without a price. The new properties of the effective Hamiltonian (1) are:

1. For the description of unbound states the effective Hamiltonian is non-Hermitian which reflects the possible leak of probability from the internal system.
2. The Hamiltonian has explicit energy dependence, making the internal dynamics highly non-linear.
3. The additional terms in the Hamiltonian that appear as a result of projection can be complicated. Even with exclusively two-body forces in the full space, the many-body interactions appear in the projected effective Hamiltonian.

By construction, the eigenvalue problem

$$\mathcal{H}(E)|\alpha\rangle = \mathcal{E}|\alpha\rangle \tag{3}$$

determines the internal part of the solution which is subject to the regular boundary condition inside matched to purely outgoing waves in the continuum. For energies E below all thresholds, the amplitudes $A_1^c(E)$ vanish, and Eq. (3) determines discrete bound states with real $\mathcal{E} = E$. Above decay thresholds, Eq. (3) has no real energy solutions, and the stationary state boundary condition can not be satisfied. The similarity of this problem to that for the bound states makes it appealing to depart from the real axis and to find discrete non-Hermitian eigenvalues. The complex energy roots $\mathcal{E}$ of (3) correspond to poles of the scattering matrix, see discussion below, and represent the many-body resonant Siegert states [8].

The transition into a complex energy plane may be rather impractical, it causes computational complications related to numerous branch cuts and unphysical roots, the relation to observables becomes complicated and rather remote. As an alternative, the Breit-Wigner approach [9] is commonly used. Here the resonances $\mathcal{E} = E - (i/2)\Gamma$ are defined as $\mathrm{Re}\,[\mathcal{E}_\alpha(E)] = E$ and $\Gamma_\alpha = -2\mathrm{Im}\,[\mathcal{E}_\alpha(E)]$. In the limit of a small imaginary part (narrow resonances) various definitions are equivalent. In the application of the CoSMo discussed below we use the Breit-Wigner approach. However, the general difficulty in parameterizing resonances in terms of centroid energies and widths should be noted. As demonstrated in Sec. 4, the problem becomes especially acute for broad resonances, high density of states, or in near-threshold situations. A look at the problem from the observable cross sections is imperative.

3. Reactions

The picture where the nuclear system is probed from "outside" is given by the transition matrix defined within the general scattering theory [2],

$$T^{ab}(E) = \sum_{12} A_1^{a*}(E) \left(\frac{1}{E - \mathcal{H}(E)} \right)_{12} A_2^b(E), \tag{4}$$

The same transition amplitudes and propagation via intrinsic space drive the process shown schematically in Fig. 1.

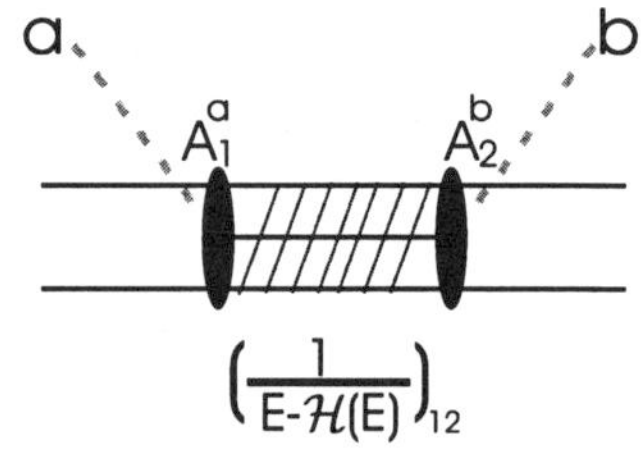

Figure 1. Reaction process: the entrance channel b with amplitude A_2^b continues through internal propagation started in the intrinsic state $|2\rangle$ driven by the non-Hermitian energy-dependent effective Hamiltonian (1) (with all excursions into continuum space included), and ends by exit from the intrinsic state $|1\rangle$ into the channel a as described by the amplitude $A_1^{a*}(E)$.

The poles of the transition matrix and related full scattering matrix $S = 1 - 2\pi i T$ are the eigenvalues of Eq. (3) located in the lower part of the complex energy plane. The reaction theory is fully consistent with resonant description in Sec. 2. However, complexity of the many-body propagator in Eq. (4) with numerous poles, interfering paths and energy dependence can make the observable cross section which is a projection of poles onto a real energy axis quite different from a collection of individual resonance peaks.

The solution of Eq. (3) with the large scale many-body Hamiltonian is a complicated task as extensively discussed in Refs. [4]. The calculation of the transition matrix (4) and of the cross section is yet another technical problem. The direct approach involving matrix inversion at all energies is extremely difficult and time consuming given large dimensions involved. The sharp resonances typically present in the spectrum make the process numerically unstable and require dense energy sampling to achieve a reasonable cross section curve. Absence of absolute numerical precision leads to instability near stable states embedded in the continuum where decays are prohibited by symmetry considerations. This problem is particularly troublesome within the m-scheme shell model approach. To overcome these difficulties, an alternative method has been developed which is discussed below.

3.1. *Unitarity and R-matrix*

The transition matrix (4) with the dimensionality equal to the number of open channels can be written as $T = \mathbf{A}^{\dagger}\mathcal{G}\mathbf{A}$, where the full effective Green's function $\mathcal{G}(E) = 1/(E - \mathcal{H})$ includes the loss of probability into all decay channels. The factorized form of the non-Hermitian part $W = 2\pi\mathbf{A}\mathbf{A}^{\dagger}$ in Eq. (1), where $\mathbf{A}$ represents a channel matrix (a set of columns of vectors A_1^c for each channel c) is the key for unitarity of the S matrix [10]. As shown in [11], the simple iteration of the Dyson equation using the definitions $\mathcal{H} = H - iW/2$ and $G = (E - H)^{-1}$ leads to the following transition and scattering matrices

$$T = \frac{R}{1 + i\pi R}, \quad S = \frac{1 - i\pi R}{1 + i\pi R}, \tag{5}$$

where the matrix $R = \mathbf{A}^{\dagger}G\mathbf{A}$ is analogous to the R-matrix of standard reaction theory; it is based on the Hermitian part of the Hamiltonian $H = H_0 + \Delta$ and computed as a function of energy using Chebyshev polynomial expansion in Sec. 3.2.

3.2. *Time evolution of the system and Green's function*

The technique behind the Green's function calculation in the CoSMo extends the idea suggested in [12] where densities of states in molecular systems were computed using the Chebyshev polynomial expansion of the time-dependent evolution operator.

First, the finite Hamiltonian matrix is rescaled and shifted by a constant so that the spectrum is mapped onto a generic energy interval [-1,1] using $H \to (H - E_+)/E_-$. The procedure involves scaling E_- and shifting E_+ parameters, $E_\pm = (E_{\max} \pm E_{\min})/2$, that are determined by the upper and lower edges of the original spectrum $E_{\max}$ and $E_{\min}$, respectively. Even within the traditional Lanczos diagonalization, the rescaling, although not required, is useful for providing numerical stability. Given a trivial nature of the rescaling procedure and its reversal, below we do not introduce special notations for the rescaled Hamiltonian.

The energy representation of the retarded propagator is given by the usual Fourier image of the evolution operator,

$$G(E) = \frac{1}{E - H} = -i \int_0^\infty dt \, \exp(iEt) \exp(-iHt), \tag{6}$$

where H is the Hamiltonian operator with a negative-definite infinitesimal imaginary part. The expansion factorizes the evolution operator using the Chebyshev polynomials as follows:

$$\exp(-iHt) = \sum_{n=0}^{\infty} (-i)^n (2 - \delta_{n0}) \, J_n(t) \, T_n(H), \tag{7}$$

where $J_n(t)$ is the usual Bessel function and the Chebyshev polynomials are defined as $T_n[\cos(\theta)] = \cos(n\theta)$. In comparison to the Taylor expansion or other methods evaluating the Green's function, the Chebyshev polynomials provide a complete set of orthogonal functions covering uniformly the interval [-1,1]. Although individual states can be resolved, the procedure is most effective when a significant energy region is involved, namely for overlapping resonances. The asymptotic of the Bessel functions assures convergence of the series. The "angle addition" equations that follow from the definition of polynomials,

$$2T_n(x)T_m(x) = T_{n+m}(x) + T_{n-m}(x) \, , \quad n \geq m, \tag{8}$$

are useful for the successive evaluation of series of vectors $|\lambda_n\rangle = T_n(H)|\lambda\rangle$ using the following iterative procedure:

$$|\lambda_0\rangle = |\lambda\rangle, \quad |\lambda_1\rangle = H|\lambda\rangle, \quad \text{and } |\lambda_{n+1}\rangle = 2H|\lambda_n\rangle - |\lambda_{n-1}\rangle. \tag{9}$$

In the CoSMo approach the calculations of reactions are performed using the Fast Fourier Transformation of Eq. (6), where the expectation value of the evolution operator in (7) is computed using iterative matrix-vector multiplications (9). In this way the R-matrix is computed which is then used to determine the cross section (5). The Chebyshev polynomials are divergent in the complex plane;

therefore only the Hermitian part $H = H_0 + \Delta$ corresponding to the R matrix can be used in the evolution operator (7). Using a conservative estimate it can be shown that n iterations lead to the energy resolution $4E_-/n$. Unlike the re-orthogonalization problem in the Lanczos algorithm, lack of numerical precision in successive matrix vector multiplication does not leads to significant deterioration of the result. $n = 1024$ was typically used for CoSMo calculations.

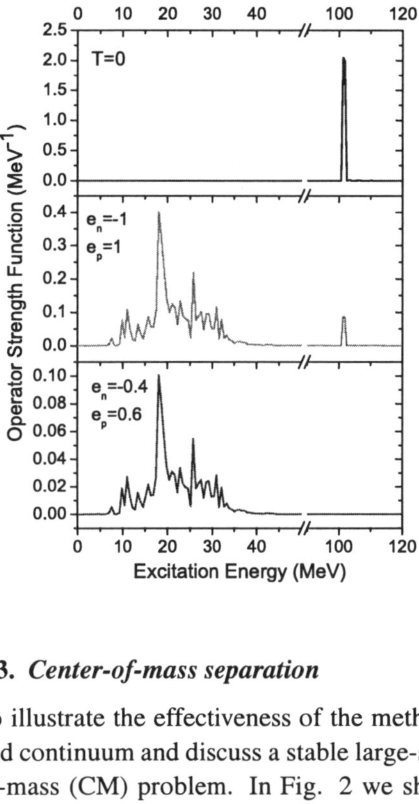

Figure 2. Strength function of a dipole operator. Upper plot: effective charges for protons and neutrons are selected equal leading to a pure isospin $T = 0$ CM operator; only the CM states with energies around 100 MeV have non-zero strength. Middle plot: effective charges are selected as -1 and +1 for neutrons and protons, respectively, the resulting mixed operator shows strength in both CM and non-CM states. Lower plot: the effective charges are selected as $e_n = -N/A = -0.4$ and $e_p = Z/A = 0.6$, which for ^{20}O excludes CM component from the dipole operator; the resulting strength shows no CM excitation.

3.3. *Center-of-mass separation*

To illustrate the effectiveness of the method, we make a digression from decays and continuum and discuss a stable large-scale shell model example of the center-of-mass (CM) problem. In Fig. 2 we show the strength function of the dipole operator. The strength function for a state $|\lambda\rangle$ is defined as

$$F_\lambda(E) = \langle\lambda|\delta(E - H)|\lambda\rangle = -\frac{1}{\pi}\,\mathrm{Im}\,\langle\lambda|G(E)|\lambda\rangle. \tag{10}$$

For the Hamiltonian H here, we consider the full $s - p - sd - pf$ shell model space with positive parity states restricted to the sd shell, while the negative parity states include all one-particle-hole excitations from the sd shell. The two-body interaction is chosen as WBP [13]. A Lawson technique is used to address the CM problem with an artificial CM vibration Hamiltonian included into H with a large scaling factor. As a result, all states that correspond to the CM excitations appear at high energy, around 100 MeV in our example. The strength function of the dipole operator $D = \sum_a e_a r_a$ is considered, where e_a is the effective charge of a particle a. In Fig. 2 the dipole strength for excitations from the 0^+ ground state of ^{20}O is plotted, namely Eq. (10) is evaluated with $|\lambda_D\rangle = D|\text{g.s}\rangle$. Depending on the choice of the effective charges for protons and neutrons, the operator D can be changed from the pure CM operator to the isovector operator containing no CM component. The change in strength of CM states is shown in Fig. 2.

4. Helium isotopes

We conclude with a realistic example of CoSMo application to the chain of helium isotopes ^{4}He to ^{10}He that serves as an illustration of all techniques combined. The internal space of this simplified model contains two single-particle levels $p_{3/2}$ and $p_{1/2}$ on top of the α-particle core. The sensitivity of the decay amplitudes A_1^c to the location of thresholds and to the parent-daughter structural relations lead to the necessity of considering the entire isotope chain. The effective shell model interactions were taken from [14,15]. These interactions are experimentally adjusted; thus it is assumed that the Hermitian renormalizations due to virtual particle excitations into continuum, $\Delta(E)$, Eq. (2), are already implicitly included; the energy dependence of $\Delta(E)$ is neglected. The diagonalization of the Hermitian many-body Hamiltonian within this valence space provides a conventional shell model solution.

One-body decays are accounted for in the model through the single-particle decay amplitudes defined as

$$A_1^c(E) = a_j(\epsilon)\ \langle 1; N|b_j^\dagger|\alpha; N-1\rangle. \tag{11}$$

These amplitudes correspond to a single particle amplitude a_j of the decay leaving a residual $N-1$ nucleon state α, while the remaining nucleons can be seen as spectators. The amplitude a_j as a function of energy is determined with the use of the Woods-Saxon potential that models the single-particle interaction between bound and continuum states. The parameters of the potential are adjusted in order to adequately represent the ^{4}He+n scattering.

The *two-body* decays can be separated into sequential and direct ones. The *sequential* decays represent higher order processes generated by the same single-particle mechanism modeled here by the Woods-Saxon potential. The *direct* decay requires introduction of new parameters describing instantaneous removal of an interacting pair. The model includes simplest two-body terms, see further discussion in [4].

In Fig. 3 the results of calculations are shown and compared to the experimental data [16,17,18]. The resonance states computed according to the Breit-Wigner definition are shown with discrete lines labeled with spin, parity and decay width. The same Fig. 3 contains a separate cross section calculation which implements techniques discussed in Secs. 3.1 and 3.2. The reaction calculation is performed with the use of the same Hamiltonian (1) and thus provides an important complementary picture to the resonant structure. The cross section shown is that for elastic neutron scattering off the ground state of the $N-1$ nucleus.

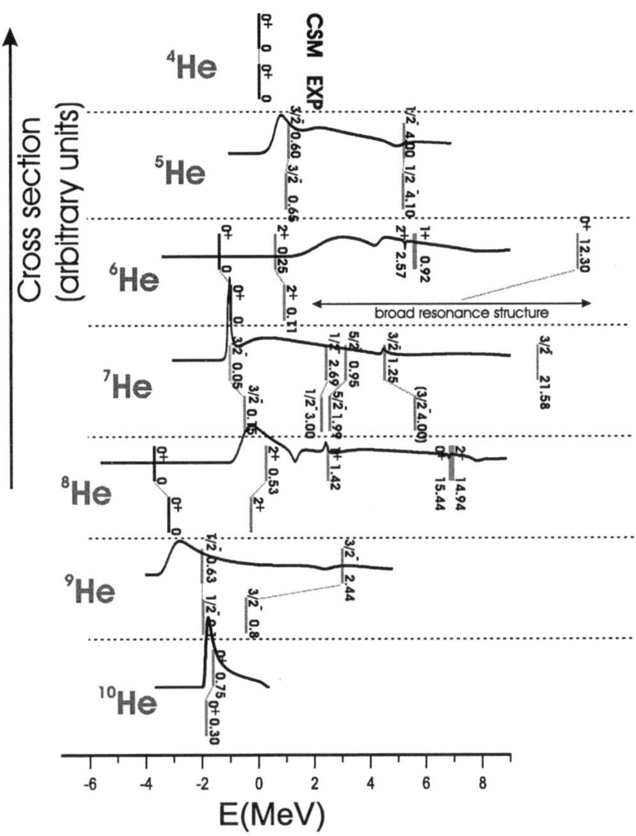

Figure 3. (Color online) CoSMo results for He isotopes. The states in the chain of isotopes starting from ^{4}He (top) up to ^{10}He (bottom) are shown as a function of the energy relative to ^{4}He. The horizontal dotted lines separate isotopes. For each case, the states from CoSMo are shown above experimentally observed states. The decay width (in units of MeV) along with spin and parity is shown for each state. The solid lines above CoSMo states show the elastic neutron scattering cross section from the spin polarized state of $N-1$ isotope in the magnetic substate with $M=0$ (even) or $M=1/2$ (odd mass) quantum number.

The model, in agreement with experiment, predicts ground states of ^{4,6,8}He to be particle-bound. The energies of these states by construction of the model exactly agree with the prediction of the traditional shell model. The states in the continuum are approached from two perspectives, via the solution of Eq. (3) under the Breit-Wigner resonance condition and by directly plotting the cross section curve. The resonance centroids are shown with discrete lines with corresponding widths indicated. The continuum coupling changes the structure of internal states leading to resonant energies being in general different from those from the shell model prediction. The resonant patterns indirectly reveal information about structure of the states and dominant decay modes. The results for ^{7}He isotope agree with recent experiments [17,18]. Our results support the "unusual structure" of the $5/2^-$ state identified by [18]. Due to its relatively high spin, this state, unlike the neighboring $1/2^-$ state, decays mainly to the 2^+ excited state in ^{6}He.

The comparison of the cross section curve with discrete resonances provides a transparent picture revealing both usefulness and limitations of the resonant parameterization approach. The scattering cross sections start from thresholds set here by the ground state of the previous $(N-1)$ nucleus. The cross sections at sharp resonances, such as the ground state of ^{7}He, agree well with the resonance parameterization. However, generally the cross section curves are not symmetric and do not show a simple, Gaussian or Lorentzian, shape. The shape of low-lying states with widths big enough to reach threshold is particularly influenced. The association of the cross section peaks with the location of the resonance is ambiguous. A remarkable example is the case of $1/2^-$ state in ^{7}He. The Breit-Wigner approach predicts an almost 3 MeV wide resonance at 2.3 MeV of excitation energy. The cross section curve, however, is only weakly influenced by such a deep pole and peaks near low energies reflecting primarily a proximity of threshold. This comparison of cross section and resonance parameterization may shed light onto the experimental controversy discussed in [17,18,19,20].

Acknowledgments

The author thanks V. Zelevinsky for collaboration. The support from the U.S. Department of Energy, grant DE-FG02-92ER40750, and National Science Foundation, grants PHY-0070911 and PHY-0244453 is acknowledged. Useful discussions with B.A. Brown and G. Rogachev are appreciated.

References

1. H. Feshbach, *Ann. Phys.* **5**, p. 357 (1958); **19**, p. 287 (1962).
2. C. Mahaux and H. Weidenmüller, *Shell-model approach to nuclear reactions* (North-Holland Pub. Co., Amsterdam, London, 1969).

3. I. Rotter, *Rep. Prog. Phys.* **54**, p. 635 (1991).
4. A. Volya and V. Zelevinsky, *Phys. Rev. C* **67**, p. 54322 (2003); *Phys. Rev. Lett.* **94**, p. 052501 (2005).
5. R. I. Betan, R. J. Liotta, N. Sandulescu and T. Vertse, *Phys. Rev. Lett.* **89**, p. 042501 (2002).
6. N. Michel, *et al.*, *Phys. Rev. Lett.* **89**, p. 042502 (2002); *Phys. Rev. C* **67**, p. 054311 (2003); *Phys. Rev. C* **70**, p. 064313 (2004).
7. J. Okolowicz, M. Ploszajczak and I. Rotter, *Phys. Rep.* **374**, p. 271 (2003).
8. A. F. J. Siegert, *Phys. Rev.* **56**, p. 750 (1939).
9. G. Breit and E. Wigner, *Phys. Rev.* **49**, p. 519 (1936).
10. L. Durand, *Phys. Rev. D* **14**, p. 3174 (1976).
11. V. Sokolov and V. Zelevinsky, *Nucl. Phys.* **A504**, p. 562 (1989).
12. T. Ikegami and S. Iwata, *J. Comput. Chem.* **23**, p. 310 (2002).
13. B. Brown, *Prog. Part. Nucl. Phys.* **47**, p. 517 (2001).
14. S. Cohen and D. Kurath, *Nucl. Phys.* **A73**, p. 1 (1965).
15. J. Stevenson, *et al.*, *Phys. Rev. C* **37**, p. 2220 (1988).
16. Evaluated nuclear structure data file http://www.nndc.bnl.gov.
17. G. V. Rogachev, *et al. Phys. Rev. Lett.* **92**, p. 232502 (2004).
18. A. A. Korsheninnikov *et al.*, *Phys. Rev. Lett.* **82**, 3581 (1999).
19. A. Wuosmaa *et al.*, *Phys. Rev. C* **72**, p. 061301(R) (2005).
20. M. Meister *et al.*, *Phys. Rev. Lett.* **88**, p. 102501 (2002).

SYMMETRY ENERGY

P. DANIELEWICZ*

National Superconducting Cyclotron Laboratory, Michigan State University, East Lansing, Michigan 48824, USA

**E-mail: danielewicz@nscl.msu.edu*

Examination of symmetry energy is carried out on the basis of an elementary binding-energy formula. Constraints are obtained on the energy value at the normal nuclear density and on the density dependence of the energy at subnormal densities.

1. Introduction

In nuclear physics, the symmetry energy is first encountered within the elementary Bethe-Weizsäcker formula for nuclear energy:

$$E = -a_V\, A + a_S\, A^{2/3} + a_C\, \frac{Z^2}{A^{1/3}} + a_A\, \frac{(N-Z)^2}{A} + \Delta\,; \tag{1}$$

it is the change in nuclear energy associated with changing neutron-proton asymmetry $(N-Z)/A$. In nuclear matter, the energy per nucleon, dependent on neutron ρ_n and proton ρ_p densities, may be represented as a sum of the energy E_0 for symmetric nuclear matter and the correction E_1 associated with the asymmetry,

$$E(\rho_n, \rho_p) = E_0(\rho) + E_1(\rho_n, \rho_p)\,, \tag{2}$$

where $\rho = \rho_n + \rho_p$. The charge symmetry of nuclear interactions, which is the symmetry under the interchange of neutrons and protons, requires that the correction be quadratic in the asymmetry, for small asymmetries:

$$E_1 = E - E_0 \simeq S(\rho) \left(\frac{\rho_n - \rho_p}{\rho} \right)^2 . \tag{3}$$

Microscopic calculations, such as [1], indicate that the quadratic approximation yields a very good representation for the energy in nuclear matter, up to the limit of neutron matter with asymmetry of 1, over a wide range of net densities. As a consequence, the energy in nuclear matter over a broad range of parameters can be described exclusively in terms of the energy of symmetric matter $E_0(\rho)$ and the symmetry-energy coefficient $S(\rho)$.

Much effort has been dedicated in the past to the understanding of $E_0(\rho)$ and less so to $S(\rho)$ whose features remain more obscure [2]. The energy $E_0(\rho)$ minimizes at the normal density ρ_0, reaching there the value of $E_0 \approx -16.0$ MeV. The uncertainties in S hamper predictions for neutron stars whose structure depends on pressure in neutron matter [3]. In pure neutron matter, the energy is $E = E_0 + S$ and the pressure is $P = \rho^2\, dE/d\rho \simeq \rho^2\, dS/d\rho$ close to ρ_0, as E_0 minimizes at ρ_0. In the calculations of neutron-star structure [3], a correlation is found, $R\,P^{-1/4} \approx$ const, between the radius R of a neutron star of a given mass and the pressure P in neutron matter at a given density $\rho \sim \rho_0$.

2. Binding Formula

When examining the standard energy formula (1), we notice that the symmetry energy has a volume character: it changes as A when the neutron and proton numbers are scaled by one factor. The formula lacks a surface symmetry term that would change as $A^{2/3}$ with the change in nucleon number. A question to ask is whether there should be such a term. Let us look at the surface energy. We can write it as $E_S = a_S\, A^{2/3} = \frac{a_S}{4\pi\, r_0^2}\, 4\pi\, r_0^2\, A^{2/3} = \frac{a_S}{4\pi\, r_0^2}\, \mathcal{S}$. The ratio $\frac{E_S}{\mathcal{S}} = \sigma = \frac{a_S}{4\pi\, r_0^2}$ is surface tension, the work that needs to be done per unit area when changing the surface area of the nucleus, such as in deforming the nucleus. The work needs to be done to compensate lost binding, as nucleons close to the surface are less bound, due to fewer neighbors, than in the interior. The formula (1) states that in the interior the nucleons are less bound in a more asymmetric nucleus. In that case, the energetic price for increasing the surface should drop, i.e. $\sigma = \frac{\partial E_S}{\partial \mathcal{S}}$ should decrease with asymmetry, in the more general definition of tension.

As intensive, σ should be expressed in terms an intensive quantity associated with the asymmetry, which is the asymmetry chemical potential:

$$\mu_A = \frac{\partial E}{\partial (N-Z)} = \frac{1}{2}\,(\mu_n - \mu_p)\,. \tag{4}$$

To lowest order, under charge symmetry, the tension needs to be quadratic in μ_A,

$$\sigma = \sigma_0 - \gamma\, \mu_A^2\,. \tag{5}$$

If the tension depends on asymmetry, so must surface energy. On examining the function $\Phi = \mu_A(N-Z) - E$, with the derivative $\partial\Phi/\partial\mu_A = N-Z$, the last dependence is seen to produce the following apparent paradox: some of the nuclear asymmetry $N-Z$ must be associated with the surface and not the interior. To answer the question on how particles can be attributed to the surface, one needs to adapt a systematic approach the separation of quantities into volume and surface contributions. Gibbs [4] proposed to consider two copies of the system, actual and an idealized reference copy where the interior densities of different quantities extend up to the surface position, cf. Fig. 1. The idealized system represents

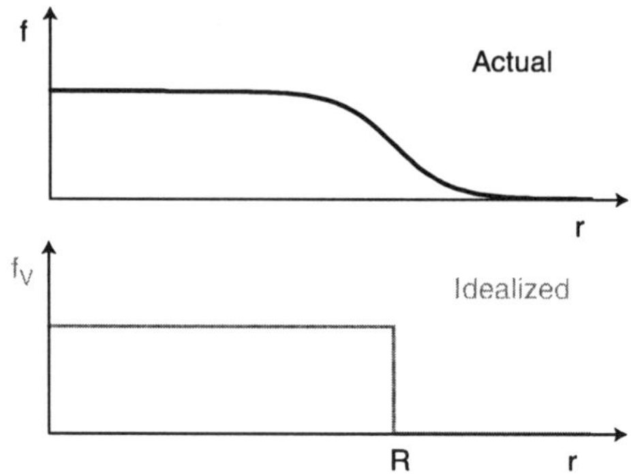

Figure 1. Gibbs' [4] construction for defining volume and surface contributions.

one with only volume contributions to different quantities, while the difference between the systems can be associated with the surface, $F_S = F - F_V$. The separation, however, depends on the position of the surface which must be set utilizing some auxiliary condition. For nuclei, it is natural to demand a vanishing surface nucleon surface number $A_S = 0$, i.e. set the surface position at the sharp-edged sphere radius R. Since nuclei, though, are binary systems, the surface positions that might be separately attributed to neutrons and protons will be, generally, displaced relative to each other, cf. Fig. 2. In consequence, in spite of a vanishing surface nucleon number, $A_S = N_S + Z_S = 0$, we may have a finite surface asymmetry $N_S - Z_S \neq 0$.

Having resolved the apparent paradox, from (5) we find

$$E_S = \sigma_0 \, \mathcal{S} + \gamma \, \mu_A^2 \, \mathcal{S} = E_S^0 + \frac{1}{4\gamma} \, \frac{(N_S - Z_S)^2}{\mathcal{S}} = E_S^0 + a_A^S \, \frac{(N_S - Z_S)^2}{A^{2/3}} \,. \quad (6)$$

Here, the index 0 refers to symmetric matter and we have introduced a surface symmetry coefficient a_A^S, with dimension of energy. For the volume energy, within the standard formula, we have

$$E_V = E_V^0 + a_A^V \, \frac{(N_V - Z_V)^2}{A} \,, \quad (7)$$

where a_A^V is the volume symmetry energy coefficient. The Coulomb energy is temporarily ignored. The net energy and asymmetry are, respectively, $E =$

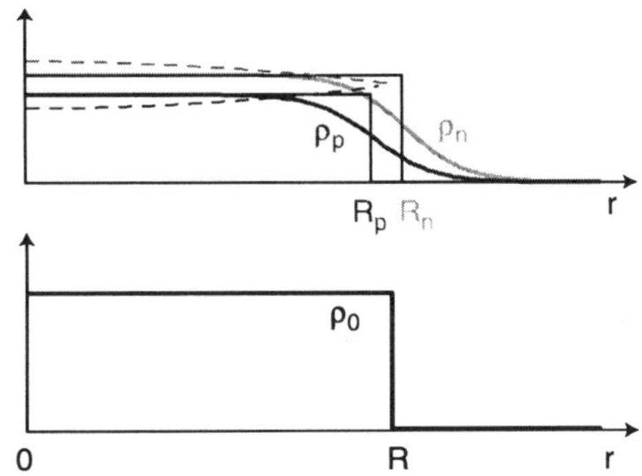

Figure 2. In a two-component system, the surfaces for the two components will be, generally, displaced relative to each other.

$E_S + E_V$ and $N - Z = N_S - Z_S + N_V - Z_V$. In the ground state, the asymmetry should partition itself into the surface and volume contributions, in such a way as to minimize the energy. The result of the energy minimization can be, actually, written right away once one notices that the surface and volume energies, quadratic in asymmetry, are analogous to the energies of capacitors quadratic in the charge. The energy of the coupled capacitors is quadratic in the net charge, with the square divided by the net capacitance, yielding:

$$E = E^0 + \frac{q^2}{2C} = E^0 + \frac{(N-Z)^2}{\frac{A}{a_A^V} + \frac{A^{2/3}}{a_A^S}} \,. \tag{8}$$

Adding now the Coulomb term, we arrive at the modified energy formula

$$E = -a_V\, A + a_S\, A^{2/3} + a_C\, \frac{Z^2}{A^{1/3}} + \frac{a_A^V}{1 + A^{-1/3}\, a_A^V/a_A^S}\, \frac{(N-Z)^2}{A} \,. \tag{9}$$

Compared to the standard formula (1), the symmetry coefficient becomes now mass dependent, $a_A(A) = a_A^V/(1 + a_A^V/(a_A^S\, A^{1/3}))$. The standard formula is recovered when $a_A^V/a_A^S = 0$, i.e. the surface does not accept asymmetry excess, or in the limit of $A \to \infty$. Within the modified formula, the symmetry coefficient weakens at low A [5,6]. Whether or not the coefficient may be replaced by the constant a_A^V depends on the ratio a_A^V/a_A^S. That ratio can be determined from the ratio of surface to volume asymmetry partitioning for the energy minimum in proportion to the capacitances:

$$\frac{N_S - Z_S}{N_V - Z_V} = \frac{C_S}{C_V} = \frac{A^{2/3}/a_A^S}{A/a_A^V} = A^{-1/3}\, \frac{a_A^V}{a_A^S} \,. \tag{10}$$

3. Asymmetry Skins

Establishing the relatively small differences in the distribution of neutrons and protons in nuclei has been, generally, difficult experimentally. Probes with different sensitivities to protons and neutrons had been utilized, such as electrons and protons, negative and positive pions, or protons and neutrons, with different associated systematic errors. The results have not been expressed in terms of the surface excess, but rather in terms of the difference in the r.m.s. radii between neutrons and protons. The conversion from the excess (10) to the difference of radii is relatively straightforward [6], if the surface diffuseness is similar for neutrons and protons. Another issue to consider theoretically is that, for heavy nuclei, Coulomb forces compete with symmetry-energy effects, pushing the proton radius out against neutrons and polarizing the nuclear interior. That competition is easily taken into account by minimizing the sum of three energies with respect to the asymmetry:

$$E = E_V + E_S + E_C \, , \tag{11}$$

where

$$E_C = \frac{e^2}{4\pi\epsilon_0} \frac{1}{R} \left(\frac{3}{5} Z_V^2 + Z_V \, Z_S + \frac{1}{2} Z_S^2 \right) . \tag{12}$$

From the energy minimization, an analytic formula for the difference of the radii follows,

$$\begin{aligned} \frac{\langle r^2 \rangle_n^{1/2} - \langle r^2 \rangle_p^{1/2}}{\langle r^2 \rangle^{1/2}} &= \frac{A}{6NZ} \frac{N-Z}{1 + A^{1/3} \, a_A^S / a_A^V} \\ &- \frac{a_C}{168 a_A^V} \frac{A^{5/3}}{N} \frac{\frac{10}{3} + A^{1/3} \, a_A^S / a_A^V}{1 + A^{1/3} \, a_A^S / a_A^V} \, . \end{aligned} \tag{13}$$

The first term on the r.h.s. represents the effects of symmetry energy only, from Eq. (10), while the second term represents the Coulomb correction. It should be mentioned that the impact of the Coulomb-symmetry energy competition is much weaker onto the net energy than onto the skin size.

Before trying to draw conclusions from data with Eq. (13), it may be worthwhile to test the macroscopic theory against the microscopic. In their nonrelativistic Hartree-Fock and relativistic Hartree calculations, Typel and Brown [7] observed correlations between the sizes of asymmetry skins in different nuclei, when utilizing different effective interactions. Those correlations are shown in Fig. 3 together with the predictions of Eq. (13) when changing the ratio a_A^S/a_A^V. The accuracy of the macroscopic theory in reproducing correlations from the microscopic theory appears to be at the level of 0.01 fm!

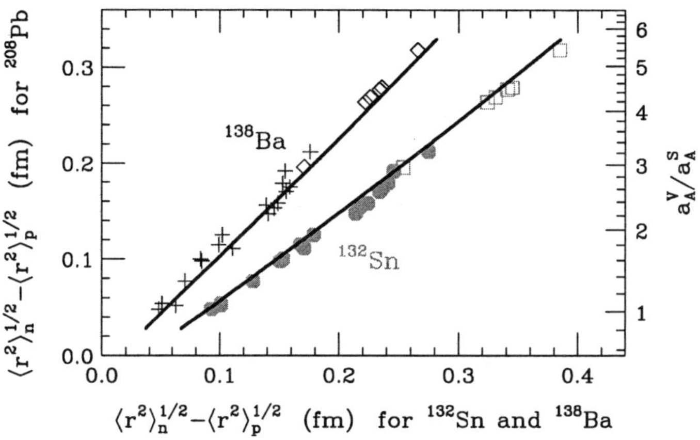

Figure 3. Correlation between the asymmetry skins for ^{208}Pb and ^{132}Sn and ^{138}Ba, in the nonrelativistic and relativistic mean-field calculations [7] (symbols) and predicted by macroscopic Eq. (13) (lines).

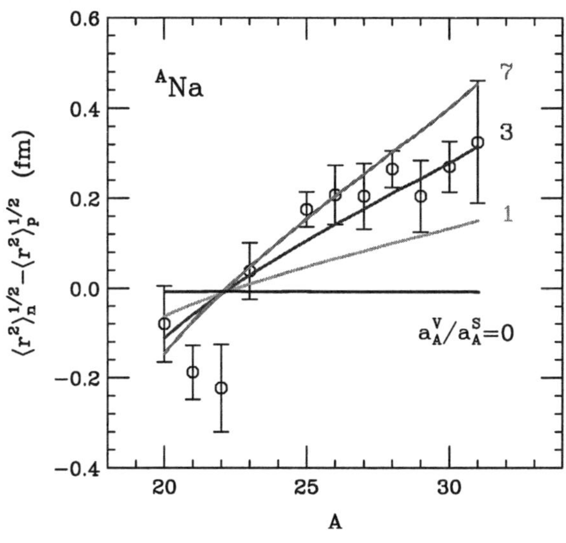

Figure 4. Asymmetry skin for Na isotopes as a function of the mass number, from the data analysis of Ref. [8] (symbols) and from Eq. (13), for the indicated values of a_A^S/a_A^V.

We next turn to the implications of skin data. Figure 4 shows a comparison of the data by Suzuki *et al.* [8] from Na isotopes to the predictions of Eq. (13) for different values of a_A^S/a_A^V. The comparison suggests a value of $a_A^S/a_A^V \sim 3$. Figure 5 shows, with sloped parallel lines, the constraints on a_A^S/a_A^V from fitting

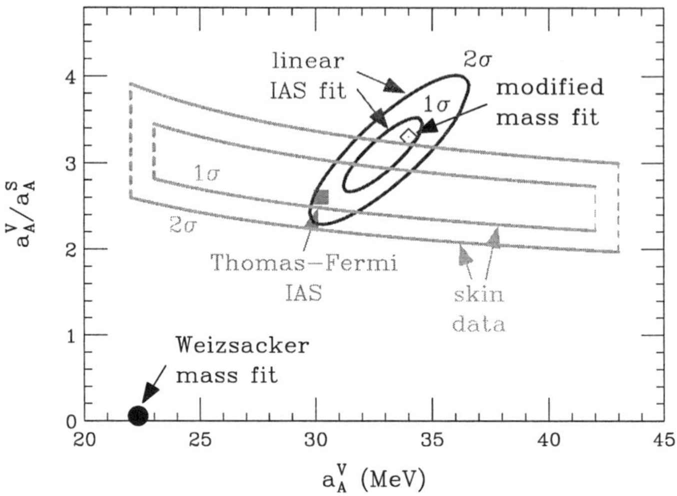

Figure 5. Constraints on the symmetry energy parameters in the plane of a_A^S/a_A^V vs a_A^V. The sloped horizontal lines represent constraints on a_A^S/a_A^V from fitting skin data at an assumed value of a_A^V. The elliptical contours represent constraints obtained from the fit linear in $A^{-1/3}$ to the values of $a_A^{-1}(A)$ from IAS.

a variety of skin data (for references to the experiments see [6]). Without Coulomb effects the constraint lines would have been horizontal; the weak sensitivity of the fits to a_A^V results from the second term on the r.h.s. of (13). The favored values of the ratio, $a_A^S/a_A^V \sim 2.8$, imply that $A^{-1/3}\, a_A^S/a_A^V$ is never small. Neither can the A-dependent symmetry coefficient be replaced by a_A^V, nor even expanded linearly in $A^{-1/3}$.

4. Isobaric Analogue States

To find absolute values of the symmetry coefficients, one might try to fit the binding formula to measured energies. However, this is treacherous as conclusions on details in different isospin-dependent terms, including Coulomb, Wigner and pairing get interrelated when drawn from a global fit to the energies. In addition, the correlation between mass number and asymmetry along the line of stability correlates the conclusions on details in the isospin dependent and isospin independent terms. The conclusions on the symmetry coefficients change depending on what is done to the other terms in the formula [6].

Optimal for determining the symmetry parameters would be a study of the symmetry term in the binding formula in isolation from the formula remainder, which might seem impossible. However, one can take advantage of the extension of charge symmetry of nuclear interactions to charge invariance. Under charge invariance the symmetry term should be a scalar in isospin space [9] and can be, thus, generalized with

$$E_A = a_A(A)\,\frac{(N-Z)^2}{A} = 4\,a_A(A)\,\frac{T_z^2}{A} \\ \rightarrow 4\,a_A(A)\,\frac{T^2}{A} = 4\,a_A(A)\,\frac{T(T+1)}{A}\,, \qquad (14)$$

where we also happen to absorb most of the Wigner term into the symmetry term. Under the generalization, the binding formula may be applied to the lowest state of a given isospin T in a nucleus. When excited, such a state is an isobaric analogue state (IAS) of the ground state of a neighboring nucleus. In the formula generalization, the pairing contribution depends on the evenness of T.

For an isospin of the same evenness as the ground, the change in the formula in the excitation occurs only in the symmetry term:

$$E_2(T_2) - E_1(T_1) = \frac{4\,a_A}{A}\{T_2(T_2+1) - T_1(T_1+1)\}\,, \qquad (15)$$

and the excitation energy can be used to the determine the symmetry energy nucleus by nucleus from

$$a_A(A) = \frac{A\,\Delta E}{4\,\Delta T^2}\,. \qquad (16)$$

In the context of the previous considerations, the question to ask is whether the deduced A-dependent symmetry coefficient weakens for light nuclei and whether the inverse of the coefficient is linear in $A^{-1/3}$:

$$a_A^{-1}(A) \stackrel{?}{=} (a_A^V)^{-1} + (a_A^S)^{-1}\,A^{-1/3}\,. \qquad (17)$$

Inverse values of symmetry coefficients, extracted according to Eq. (15) from IAS data [10], are shown in Fig. 6. It is seen that the inverse coefficient changes with $A^{-1/3}$ in a roughly linear fashion, although significant shell effects are present. The line across the figure represents best fit with Eq. (17). The 1- and 2-σ constraints on the symmetry coefficients from the fit are further indicated with elliptical contours in Fig. 5. Combining the constraints from the fits, we conclude that $30.0\,\mathrm{MeV} \lesssim a_A^V \lesssim 32.5\,\mathrm{MeV}$ and $2.6 \lesssim a_A^V/a_A^S \lesssim 3.0$.

5. Consequences and Conclusion

The emergence of the surface capacitance for asymmetry may be tied to the weakening of the symmetry energy with density, see e.g. [11,6]. Due to the weakening, it becomes advantageous for the nucleus to push its asymmetry to the surface to

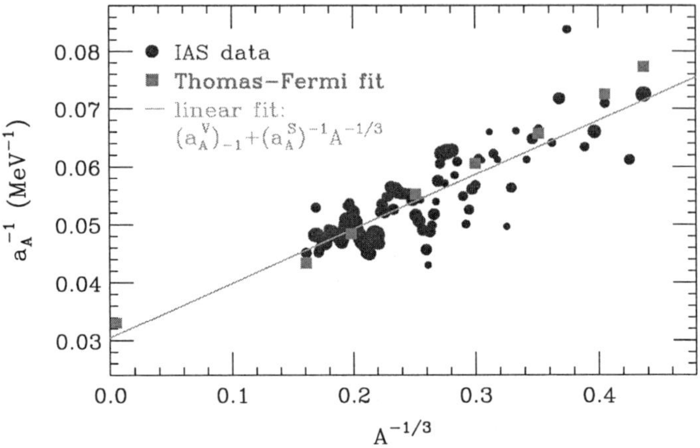

Figure 6. Inverse of the A-dependent symmetry coefficient as a function of $A^{-1/3}$. Circles represent values extracted with (15) from extremal IAS excitation energies in [10]. Circle size is proportional to the factor $\Delta T^2/A$ in the coefficient determination. The line and squares show results of the fits to the experimental results following either Eq. (17) or Thomas-Fermi theory [13].

lower energy. The ratio of the symmetry coefficients, specifically, can be tied to the shape of the symmetry energy dependence on density as, in the local-density approximation to the symmetry energy, the ratio is found to be

$$\frac{a_A^V}{a_A^S} = \frac{3}{r_0} \int dr \, \frac{\rho(r)}{\rho_0} \left[\frac{S(\rho_0)}{S(\rho(r))} - 1 \right] . \tag{18}$$

Here, the integration is across the nuclear surface and $\rho(r)$ is the density as a function of position. For density-independent symmetry energy, $S(\rho) = S(\rho_0) \equiv a_A^V$, the surface does not accept the asymmetry, $a_A^V/a_A^S = 0$!

Using the correlations between the coefficient ratio, skins and drop of the symmetry energy with ρ, within the relativistic and nonrelativistic calculations by Fuhrnstahl [12], one can arrive at limits at on the drop, either expressed in terms of the value of symmetry energy at half of the normal density or in terms of the power of density in parameterization of the symmetry energy [13]. Specifically, one finds $0.58 \lesssim S(\rho_0/2)/a_A^V \lesssim 0.69$ and $0.54 \lesssim \gamma \lesssim 0.77$ in $S(\rho) \simeq a_A^V(\rho/\rho_0)^\gamma$. These further imply limits on pressure in neutron matter at normal density and, with results of Ref. [3], produce limits on neutron-star radius of $11.5\,\text{km} \lesssim R \lesssim 13.5\,\text{km}$ for $1.4\,M_\odot$ mass.

To conclude, the requirement of macroscopic consistency brings in the surface symmetry energy into the nuclear binding formula. The volume and surface symmetry energies combine as energies of coupled capacitors. The extension of the binding formula implies emergence of the asymmetry skins for nuclei and weakening of the symmetry term in light nuclei. The systematic of the asymmetry skins restricts ratio of the symmetry coefficients. The charge invariance allows to study variation of the symmetry coefficient nucleus by nucleus. Combination of the fits to skins and IAS yields $30.0\,\text{MeV} \lesssim a_A^V \lesssim 32.5\,\text{MeV}$ and $2.6 \lesssim a_A^V/a_A^S \lesssim 3.0$. The surface symmetry energy is associated with weakening of the symmetry energy with density. The a_A^V/a_A^S ratio implies limits on drop characteristics, such as $0.58 \lesssim S(\rho_0/2)/a_A^V \lesssim 0.69$. Implications for neutron stars follow. Current direction is to incorporate shell corrections into the IAS analysis.

References

1. I. Bombaci and U. Lombardo, Phys. Rev. C 44, 1892 (1991).
2. B. A. Brown, Phys. Rev. Lett. 85, 5296 (2000).
3. J. M. Lattimer and M. Prakash, Astrophys. J. 550, 426 (2001).
4. J. W. Gibbs, *The Collected Works* (Yale Univ. Press, New Haven, 1948).
5. W. D. Myers and W. J. Swiatecki, Ann. Phys. 84, 186 (1974).
6. P. Danielewicz, Nucl. Phys. A 727, 233 (2003).
7. S. Typel and B. A. Brown, Phys. Rev. C 64, 027302 (2001).
8. T. Suzuki *et al.*, Phys. Rev. Lett. 75, 3241 (1995).
9. J. Jänecke, T. W. O'Donnell and V. I. Goldanskii, Nucl. Phys. A 728, 23 (2003).
10. M. S. Antony, A. Pape and J. Britz, Atomic Data and Nuclear Data Tables 66, 1 (1997).
11. A. R. Bodmer, Nucl. Phys. 9, 371 (1958/1959).
12. R. J. Furnstahl, Nucl. Phys. A 706, 85 (2002).
13. P. Danielewicz, nucl-th/0411115.

TENSOR INTERACTIONS IN MEAN-FIELD APPROACHES

J. DOBACZEWSKI
Institute of Theoretical Physics, Warsaw University
ul. Hoża 69, 00-681 Warsaw, Poland
E-mail: Jacek.Dobaczewski@fuw.edu.pl

Basic properties of the nuclear tensor mean fields are reviewed, and their role in changing the shell structure and masses of nuclei is analyzed within the spherical Hartree-Fock-Bogolyubov approach.

1. Introduction

Modern formulation of the nuclear mean-field theory is based on the energy density formalism,[1] which has been over the years developed for electronic systems. According to the formal Hohenberg-Kohn[2] and Kohn-Sham theorems,[3] exact ground-state energies of many-fermion systems can be obtained by minimizing certain exact functional of one-body density. These theorems do not provide any method to construct the exact functional in a systematic way; nevertheless one can build phenomenological functionals and test their performance against experimental data. Such an approach is also consistent with the ideas of the effective field theory, whereupon properties of composite objects at low energies can be described by Lagrangians which include high-energy dynamics in the form of the appropriate series of contact terms.

Within the energy density formalism, one treats the nucleus as a single composite object described by a set of one-body densities. At low energies, when the densities are varying slowly in the nuclear interior and then go smoothly to zero at the nuclear surface, one can consider only local densities that are built of the one-body density matrix and its derivatives up to the second-order. Systematics construction of the most general energy density functional (EDF) consistent with symmetries is then possible,[4] and gives a generalization of the extremely successful approach based on the Skyrme effective interaction.[5]

The origins of the spin-orbit (SO) splitting in nuclei can be attributed to the bare two-body SO and tensor interactions,[6] which contribute differently to the spin-saturated (SS) and spin-unsaturated (SUS) nuclei. An alternative explanation is also sought in the relativistic mean-field theories with meson couplings[7], where no distinction between the SS and SUS systems is obtained.

The Skyrme interaction was introduced into nuclear physics more then 30 years ago,[8] and shortly after it was supplemented by tensor forces.[9,10] However, after these ground-breaking studies, in most of the subsequent applications the tensor forces were not taken into account. Moreover, within many Skyrme-force parameterizations constructed to date, the tensor terms in the EDF that were coming from the central force, were quite arbitrarily set equal to zero.[5]

In the present paper, I discuss the form of the tensor terms in the EDF and their influence on the single-particle energies and nuclear masses. In Sec. 2, I recall the recent experimental evidence on the changes of shell structure in neutron-rich $Z\approx20$ nuclei.[11,12] This is only one of several such examples recently identified in light nuclei and interpreted within the shell-model by introducing tensor interactions.[13,14] Properties of the tensor terms in the EDF are discussed in Sec. 3, and in Sec. 4, I present results of calculations for single-particle energies and masses obtained with tensor terms included in the mean-field approach.

2. Shell structure of neutron-rich $Z\approx20$ nuclei

In a series of recent experiments performed at the Argonne National Laboratory with Gammasphere[11] and National Superconducting Cyclotron Laboratory,[12] properties of low-lying collective states of even-even neutron-rich titanium isotopes have been measured. As illustrated in Fig. 1, the data reveal the presence of a closed N=32 subshell, in addition to the standard N=28 shell present in heavier elements. Indeed, both ^{50}Ti and ^{54}Ti show the increased 2^+ energies, and decreased BE2 values, as compared to their neighbours.

In order to explain such a change of the shell structure, the single-particle neutron $\nu f_{5/2}$ orbital must be shifted up, which leaves a gap between the spin-orbit-split $\nu p_{3/2}$ and $\nu p_{1/2}$ orbitals, and creates a subshell closure for four particles occupying $\nu p_{3/2}$. The shell-model calculations,[15] performed for the single-particle orbitals shifted in this way, confirm the pattern shown in Fig. 1. The origins of the shift are attributed to the decreased monopole interaction energy between the proton $\pi f_{7/2}$ and neutron $\nu f_{5/2}$ orbitals, which occurs when protons are removed from $\pi f_{7/2}$. The source of such a monopole interaction is in turn attributed to the shell-model tensor interaction between these orbitals.

Positions of single-particle levels can be best studied within the mean-field approximation, in which they are basic dynamic characteristics of the system, resulting from the two-body interactions being averaged with particle densities of occupied states. Therefore, in this paper I evaluate the single-particle energies by

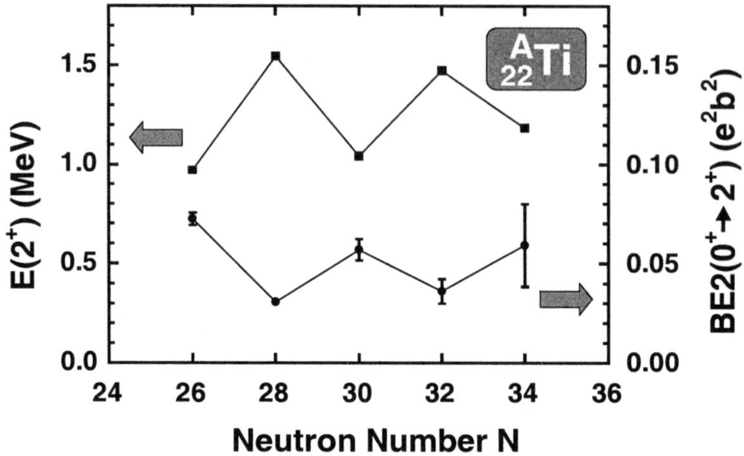

Figure 1. Energies of the first excited 2^+ states (squares, left scale) and reduced transition probabilities BE2($0^+ \to 2^+$) (circles, right scale), measured in neutron-rich Ti isotopes.[11,12]

applying the mean-field methods to tensor interactions.

3. Tensor densities in the energy density functional

3.1. *Spin-orbit and tensor forces*

Momentum-dependent two-body SO[8] and tensor[9,10] interactions have the form

$$\begin{aligned} \hat{V}_{SO} &= iW_0 \hat{\boldsymbol{S}} \cdot \left[\hat{\boldsymbol{k}}' \times \hat{\boldsymbol{k}}\right], \\ \hat{V}_{Te} &= \tfrac{1}{2} t_e \left[\hat{\boldsymbol{k}}' \cdot \hat{\mathsf{S}} \cdot \hat{\boldsymbol{k}}' + \hat{\boldsymbol{k}} \cdot \hat{\mathsf{S}} \cdot \hat{\boldsymbol{k}}\right], \\ \hat{V}_{To} &= t_o \hat{\boldsymbol{k}}' \cdot \hat{\mathsf{S}} \cdot \hat{\boldsymbol{k}}, \end{aligned} \tag{1}$$

where the vector and tensor spin operators read

$$\begin{aligned} \hat{\boldsymbol{S}} &= \boldsymbol{\sigma}_1 + \boldsymbol{\sigma}_2, \\ \hat{\mathsf{S}}^{ij} &= \tfrac{3}{2}\left[\boldsymbol{\sigma}_1^i \boldsymbol{\sigma}_2^j + \boldsymbol{\sigma}_1^j \boldsymbol{\sigma}_2^i\right] - \delta^{ij} \boldsymbol{\sigma}_1 \cdot \boldsymbol{\sigma}_2. \end{aligned} \tag{2}$$

When averaged with one-body density matrices, these interactions contribute to the following terms in the EDF (see Refs.[16,4] for derivations),

$$\begin{aligned} \mathcal{H}_{SO} &= \tfrac{1}{4} W_0 \left[3 J_0 \tfrac{d\rho_0}{dr} + J_1 \tfrac{d\rho_1}{dr}\right], \\ \mathcal{H}_T &= \tfrac{5}{8}\left[t_e J_n J_p + t_o (J_0^2 - J_n J_p)\right], \end{aligned} \tag{3}$$

where the conservation of time-reversal and spherical symmetries was assumed. Here, ρ_t and J_t are the neutron, proton, isoscalar, and isovector particle and SO densities[8,16,4] for t=n, p, 0, and 1, respectively.

Apart from the contribution of the SO energy density to the central potential, variation of the SO and tensor terms with respect to the densities yields the one-body SO potential for neutrons (t=n) and protons (t=p),

$$W_t^{SO} = \frac{1}{2r}\left[W_0\left(\frac{d\rho_0}{dr}+\frac{d\rho_t}{dr}\right)+\frac{5}{8}\Big((t_e+t_o)J_0-(t_e-t_o)J_t\Big)\right]\hat{\boldsymbol{L}}\cdot\hat{\boldsymbol{S}}. \quad (4)$$

Hence, it is clear that the only effect of including the tensor interaction is a modification of the SO splitting of the single-particle levels, and that, from the point of view of one-body properties, tensor interactions act very similarly to the two-body SO interactions. However, the latter ones induce the SO splitting that is only weakly depending on the shell filling. This is so because the corresponding form-factor in Eq. (4) is given by the radial derivatives of densities. On the other hand, the SO splitting induced by the tensor forces depends strongly on the shell filling, because its form-factor is given by the SO densities $J(r)$. Indeed, when only one of the SO partners is occupied (SUS system), the SO density is large, and when both partners are occupied (SS system), the SO density is small, see Sec. 4 for numerical examples.

3.2. *Spin-orbit and tensor energy densities*

Within the energy-density approach, one does not relate the EDF to an average of the two-body force, but one postulates the EDF based on symmetry conditions only. Then, the most general EDF, depending on the spin-current densities, reads[4]

$$\begin{aligned}\mathcal{H}_{SO} &= \textstyle\sum_{t=0,1} C_t^{\nabla J}\rho_t\boldsymbol{\nabla}\cdot\boldsymbol{J}_t,\\ \mathcal{H}_{T} &= \textstyle\sum_{t=0,1}\left(C_t^{J0}\underline{J}_t^2+C_t^{J1}\boldsymbol{J}_t^2+C_t^{J2}\underline{\mathsf{J}}_t^2\right),\end{aligned} \quad (5)$$

where the spherical-symmetry condition has been released. The standard pseudoscalar $\underline{J}_t$, vector $\boldsymbol{J}_t$, and pseudotensor $\underline{\mathsf{J}}_t$ parts of the spin-current density,[16]

$$J_{abt} = \frac{1}{2i}\Big[(\boldsymbol{\nabla}-\boldsymbol{\nabla}')_a s_{bt}(\boldsymbol{r},\boldsymbol{r}')\Big]_{\boldsymbol{r}=\boldsymbol{r}'}, \quad (6)$$

are defined as

$$\begin{aligned}\underline{J}_t &= \textstyle\sum_{a=x,y,z} J_{aat},\\ \boldsymbol{J}_{at} &= \textstyle\sum_{b,c=x,y,z}\epsilon_{abc}J_{bct},\\ \underline{\mathsf{J}}_{abt} &= \frac{1}{2}J_{abt}+\frac{1}{2}J_{bat}-\frac{1}{3}\underline{J}_t\delta_{ab}.\end{aligned} \quad (7)$$

The tensor energy density $\mathcal{H}_T$ now depends on six coupling constants, C_t^{J0}, C_t^{J1}, and C_t^{J2}, for t=0,1, and not on two coupling constants, t_e and t_o, as in Eq. 3. Similarly, the SO energy density $\mathcal{H}_{SO}$ depends now on two coupling constants

$C_t^{\nabla J}$ for t=0,1, and not on one, W_0, (the latter generalization has been introduced and studied in Ref.[17]).

From the symmetry conditions imposed by the spherical, axial, and reflection symmetries one obtains[18] that:

1° The pseudoscalar densities $\underline{J}_t$ vanish unless the axial or reflection symmetries are broken.

2° The pseudotensor densities $\underline{\mathsf{J}}_t$ vanish unless the spherical symmetry is broken.

Finally, the gauge-invariance symmetry conditions[4] require that there are only two gauge-invariant combinations of the pseudoscalar, vector, and pseudotensor terms, namely,

$$\begin{aligned} G_t^T &= \tfrac{1}{3}\underline{J}_t^2 + \tfrac{1}{2}\boldsymbol{J}_t^2 + \underline{\mathsf{J}}_t^2, \\ G_t^F &= \tfrac{2}{3}\underline{J}_t^2 - \tfrac{1}{4}\boldsymbol{J}_t^2 + \tfrac{1}{2}\underline{\mathsf{J}}_t^2. \end{aligned} \tag{8}$$

In such a case, only four out of the six tensor coupling constants are linearly independent, i.e.,

$$\begin{aligned} C_t^{J0} &= \tfrac{1}{3}A_t + \tfrac{2}{3}B_t, \\ C_t^{J1} &= \tfrac{1}{2}A_t - \tfrac{1}{4}B_t, \\ C_t^{J2} &= \; A_t + \tfrac{1}{2}B_t. \end{aligned} \tag{9}$$

On the other hand, the averaging of the tensor forces (1) implies that only two out of the six tensor coupling constants remain linearly independent, i.e.,

$$\begin{aligned} B_0 = -3A_0 &= -\tfrac{3}{8}\left(t_e + 3t_o\right), \\ B_1 = -3A_1 &= \;\;\tfrac{3}{8}\left(t_e - t_o\right). \end{aligned} \tag{10}$$

4. Mean-field calculations with tensor terms included in the energy density functional

4.1. *Spin-orbit densities*

In order to illustrate the influence of tensor densities on single-particle and global nuclear properties, in Fig. 2 are shown the radial components, $J_n(r)$=$\boldsymbol{J}_n \cdot \boldsymbol{r}/r$, of the neutron vector SO densities $\boldsymbol{J}_n$, calculated for the nickel isotopes between N=28 and 50. Calculations have been performed for the Skyrme SLy4 interaction,[19] by using the Hartree-Fock-Bogolyubov (HFB) method with the spherical symmetry assumed.[20]

One can see that the SO densities are mostly positive and peaked near the surface. At N=28, the SO density is large and its major part comes from the occupied

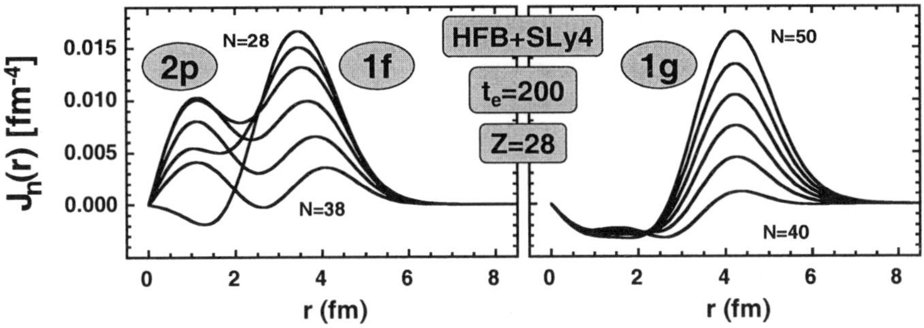

Figure 2. Radial components $J_n(r)$ of the neutron vector SO densities $\boldsymbol{J}_n$, calculated for the Z=28 isotopes with N=28–38 (left panel) and N=40–50 (right panel).

$\nu f_{7/2}$ orbital (SUS system). By adding neutrons in the shell above N=28, this part is gradually cancelled by an increasing in magnitude, negative contribution from the SO partner $\nu f_{5/2}$. At the same time contributions from the $\nu p_{3/2}$ and $\nu p_{1/2}$ orbitals appear. When both pairs of the SO partners are occupied around N=38, and when the $\nu g_{9/2}$ orbital is still empty (SS system), the SO density is rather small. Beyond N=40, it increases again until the $\nu g_{9/2}$ orbital becomes fully occupied at N=50. Note the shift of the SO densities to larger distances, which occurs at the point of the switch-over between the dominating $1f$ and $1g$ contributions.

A similar pattern of varying SO densities is valid for all shells. For SS systems, one obtains small SO densities, while for SUS systems, the SO densities are large. Therefore, the SO densities are small at magic shells N, Z=2, 8, and 20, and large at magic shells N, Z=28, 50, 82, and 126.

Since the $j_>$ partners are always occupied first, the SO densities are mostly positive. Note that the derivatives of particle densities are mostly negative, and also peaked at the surface; therefore, for positive coupling constants, the SO and tensor forces split the SO partners in opposite directions, cf. Eq. (4).

4.2. *Single-particle levels*

In Fig. 3 are shown properties of neutron single-particle levels calculated for the chain of the N=32 isotones (these levels are relevant for the changes of the shell structure discussed in Sec. 2). Single-particle energies $\varepsilon_N^{n\ell j}$ (top panels) were calculated as the canonical energies[21] of the HFB method. In order to better visualize the dependence of the single-particle energies on the proton number, in the middle and bottom panels are shown the SO splittings and centroids, respectively, of the

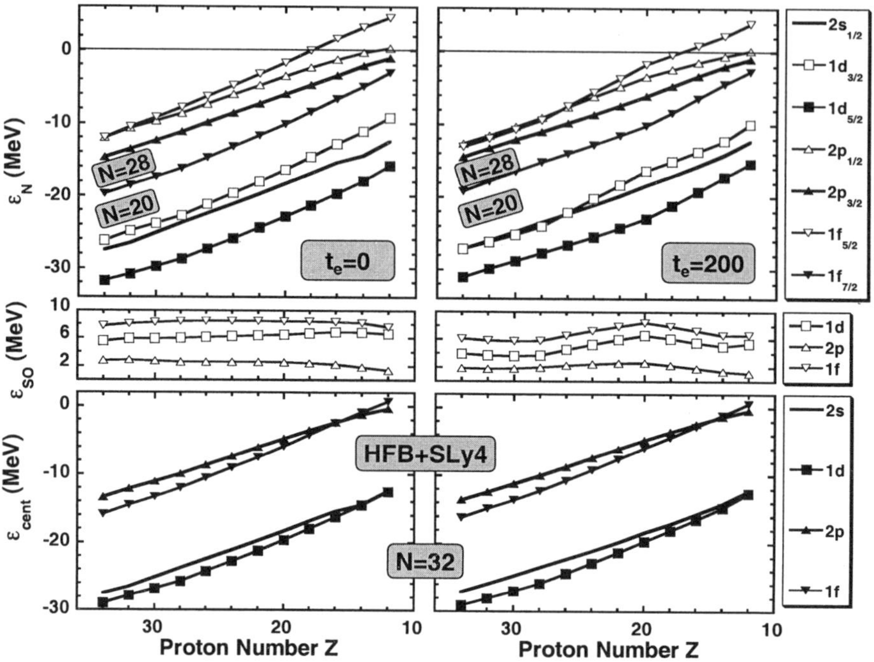

Figure 3. Single-particle properties of neutron levels in N=32 isotones, calculated within the HFB method with the SLy4 Skyrme interaction. Left panels correspond to the standard SLy4 parametrization with no tensor terms (t_e=0) while the right panels correspond to the tensor-even interaction included (t_e=200 MeV fm^5). Single-particle energies in the $sdpf$ shell (top panels) are shown along with their spin-orbit-splitting energies (middle panels) and centroid energies (bottom panels).

SO partners, defined as

$$\begin{aligned} \varepsilon_{\rm SO}^{n\ell} &= \quad \varepsilon_N^{n\ell j_<} - \varepsilon_N^{n\ell j_>}, \\ \varepsilon_{\rm cent}^{n\ell} &= \tfrac{1}{2}\left(\varepsilon_N^{n\ell j_<} + \varepsilon_N^{n\ell j_>}\right). \end{aligned} \tag{11}$$

Without the tensor terms (left panels of Fig. 3), the neutron single-particle energies vary smoothly with the proton numbers. Apart from a gradual increase with decreasing Z, one observes two clear type of changes in the shell structure. First, in each shell the centroids of levels with different values of ℓ become degenerate towards the neutron drip line. This effect is related to the increase of the surface diffuseness of particle distributions, which renders the shell structure of very neutron-rich nuclei similar to that of a harmonic oscillator.[22] Second, near the neutron drip line the SO splitting of the weakly bound p orbitals becomes smaller, because such orbitals start to decouple from the SO potential due to their increasing spatial dimensions.[23]

In the right panels of Fig. 3 are shown the analogous results obtained with the tensor-even interaction $\hat{V}_{Te}$, Eq. (1), taken into account for t_e=200 MeV fm^5. (This particular value of the coupling constant was not optimized in any sense, and it is used here only to illustrate some general trends.) The use of the tensor-even interaction (T=0, S=1 neutron-proton channel) corresponds to the energy density that depends on the product of neutron and proton SO densities, cf. Eq. (3). Therefore, the effect of the tensor term vanishes at the closed proton shell Z=20 (SS system). For Z higher (lower) than 20, the effect of the tensor term increases as a result of increasing contributions to the proton SO density coming from the $\pi f_{7/2}$ ($\pi d_{3/2}$) orbitals. This is clearly visible in the middle right panel of Fig. 3, where the SO splitting decreases on both sides of ^{52}Ca. This is so because, for positive values of the coupling constant t_e, the effect of the tensor force partly cancels that of the standard SO force. As a result, the $\pi f_{5/2}$ orbital is in ^{60}Ni much closer to the p orbitals than it is in ^{54}Ti; the shift which is compatible with the changes of the shell structure discussed in Sec. 2.

A detailed reproduction of the level positions is not the goal of the present study. The coupling constants of the tensor terms have to be adjusted together with other parameters of the EDF, by considering not only this particular region of nuclei, and not only this particular set of observables. Indeed, the tensor terms included in the EDF will influence many different global nuclear properties throughout the mass chart, and a global analysis is therefore necessary. Before this is done, in the next section, the impact of the tensor interaction on nuclear binding energies is studied in a preliminary way.

4.3. *Binding energies*

When the tensor terms (3) are added to the EDF, the binding energies are affected through self-consistent changes of all the terms in the EDF. However, qualitatively, the effects of tensor terms on the ground-state energies can be illustrated by integrals of products of the SO densities that appear in Eq. (3). In Fig. 4, values of such integrals are shown for the neutron SO densities squared, $J_n^2(r)$, calculated at magic proton numbers in function of the neutron numbers. Comparison of results obtained without (left panel) and with (right panel) tensor-even interaction included, shows that the effect of the tensor term can, in the first approximation, be treated perturbatively. Due to the fact that the proton SO densities depend weakly on the neutron numbers, for Z=28, 50, and 82 the integrals of products $J_n(r)J_p(r)$ show similar a behaviour to those of $J_n^2(r)$, while they are small for Z=8 and 20.

From the results shown in Fig. 4, it is clear that, for positive coupling constants, the tensor terms will give characteristic contributions to the ground-state energies of heavy nuclei. These contributions will have a form of inverted arches, spanned between the neutron magic numbers. This feature is conspicuously remi-

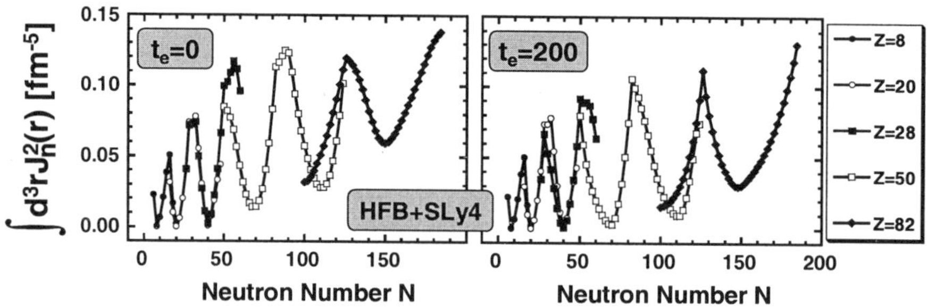

Figure 4. Integrals of the neutron SO densities squared that define contributions of the tensor terms to total binding energies. Left panels correspond to the standard SLy4 parametrization with no tensor terms (t_e=0), while the right panels correspond to the tensor-even interaction included (t_e=200 MeV fm^5).

5. Conclusions

During the past thirty odd years, when the nuclear self-consistent mean-field methods based on effective interactions were developed and implemented, the tensor interactions have been largely ignored. On the other hand, recent experimental studies and shell-model analyses indicate that these interactions may play an important role in several regions of nuclear chart. A unified picture of the role played by tensor interactions throughout the mass table is still missing, and is very much needed. In the present paper, I have reviewed basic properties of the tensor mean fields, and I have illustrated their role in changing the shell structure and masses of nuclei.

Acknowledgments

This work was supported in part by the Polish Committee for Scientific Research (KBN) under contract N0. 1 P03B 059 27 and by the Foundation for Polish Science (FNP).

References

1. R.M. Dreizler and E.K.U. Gross, *Density Functional Theory* (Springer, Berlin, 1990).
2. P. Hohenberg and W. Kohn, Phys. Rev. 136, B864 (1964).
3. W. Kohn and L.J. Sham, Phys. Rev. **140**, A1133 (1965).
4. E. Perlińska, S.G. Rohoziński, J. Dobaczewski, and W. Nazarewicz, Phys. Rev. C **69**, 014316 (2004).
5. M. Bender, P.-H. Heenen, and P.-G. Reinhard, Rev. Mod. Phys. **75**, 121 (2003).
6. R.R. Scheerbaum, Phys. Lett. **B61**, 151 (1976); **B63**, 381 (1976); Nucl. Phys. **A257**, 77 (1976).
7. P. Ring, Prog. Part. Nucl. Phys. **37**, 193 (1996).
8. D. Vautherin and D.M. Brink, Phys. Rev. **C5**, 626 (1972).
9. H. Flocard, Thesis 1975, Orsay, Série A, N° 1543.
10. Fl. Stancu, D.M. Brink, H. Flocard, Phys. Lett. **68B**, 108 (1977).
11. B. Fornal, S. Zhu, R.V.F. Janssens, M. Honma, R. Broda, P.F. Mantica, B.A. Brown, M.P. Carpenter, P.J. Daly, S.J. Freeman, Z.W. Grabowski, N.J. Hammond, F.G. Kondev, W. Krolas, T. Lauritsen, S.N. Liddick, C.J. Lister, E.F. Moore, T. Otsuka, T. Pawlat, D. Seweryniak, B.E. Tomlin, and J. Wrzesinski, Phys. Rev. C **70**, 064304 (2004).
12. D.-C. Dinca, R.V.F. Janssens, A. Gade, D. Bazin, R. Broda, B.A. Brown, C.M. Campbell, M.P. Carpenter, P. Chowdhury, J.M. Cook, A.N. Deacon, B. Fornal, S.J. Freeman, T. Glasmacher, M. Honma, F.G. Kondev, J.-L. Lecouey, S.N. Liddick, P.F. Mantica, W.F. Mueller, H. Olliver, T. Otsuka, J.R. Terry, B.A. Tomlin, and K. Yoneda, Phys. Rev. C **71**, 041302 (2005).
13. T. Otsuka, R. Fujimoto, Y. Utsuno, B.A. Brown, M. Honma, and T. Mizusaki, Phys. Rev. Lett. **87**, 082502 (2001).
14. T. Otsuka, T. Suzuki, R. Fujimoto, H. Grawe, and Y. Akaish, Phys. Rev. Lett. **95**, 232502 (2005).
15. M. Honma, T. Otsuka, B.A. Brown, and T. Mizusaki, Eur. Phys. Jour. **25,s01**, 499 (2005).
16. Y.M. Engel, D.M. Brink, K. Goeke, S.J. Krieger, and D. Vautherin, Nucl. Phys. **A249**, 215 (1975).
17. P.-G. Reinhard and H. Flocard, Nucl. Phys. **A584**, 467 (1995).
18. S.G. Rohoziński, J. Dobaczewski, and W. Nazarewicz, to be published.
19. E. Chabanat, P. Bonche, P. Haensel, J. Meyer, and F. Schaeffer, Nucl. Phys. **A635**, 231 (1998).
20. K. Bennaceur and J. Dobaczewski, Comput. Phys. Commun. **168**, 96 (2005).
21. J. Dobaczewski, W. Nazarewicz, T.R. Werner, J.-F. Berger, C.R. Chinn, and J. Dechargé, Phys. Rev. **C53**, 2809 (1996).
22. J. Dobaczewski, I. Hamamoto, W. Nazarewicz, and J.A. Sheikh, Phys. Rev. Lett. **72**, 981 (1994).
23. W. Pöschl, D. Vretenar, G.A. Lalazissis, and P. Ring, Phys. Rev. Lett. **79**, 3841 (1977).
24. J. Dobaczewski, M. Stoitsov, and W. Nazarewicz, Proc. Int. Conf. on *NUCLEAR PHYSICS, LARGE and SMALL, Microscopic Studies of Collective Phenomena*, Cocoyoc, México, April 19-22, 2004, AIP Conference Proceedings Volume 726, ed. R. Bijker, R.F. Casten, and A. Frank (American Institute of Physics, New York, 2004)

p. 51, nucl-th/0404077.
25. M. Bender, G. Bertsch, and P.-H. Heenen, Phys. Rev. Lett **94**, 102503 (2005).
26. M. Bender, G. Bertsch, and P.-H. Heenen, Phys. Rev. C **73**, 034322 (2006).

ASTROPHYSICAL CHALLENGES TO RIA: EXPLOSIVE NUCLEOSYNTHESIS IN SUPERNOVAE

G. MARTÍNEZ-PINEDO, A. KELIĆ, K. LANGANKE, K.-H. SCHMIDT

Gesellschaft für Schwerionenforschung,
D-64291 Darmstadt, Germany

D. MOCELJ, C. FRÖHLICH, F.-K. THIELEMANN, I. PANOV, T. RAUSCHER, M. LIEBENDÖRFER

Department of Physics and Astronomy, University of Basel
Klingelbergstrasse 82, CH-4056 Basel, Switzerland

N. T. ZINNER

Institute for Physics and Astronomy, University of Århus,
DK-8000 Århus C, Denmark

B. PFEIFFER

Institute for Nuclear Chemistry, University of Mainz
Fritz-Strassmann-Weg 2, D-55128 Mainz, Germany

R. BURAS AND H.-TH. JANKA

Max-Planc-Institut für Astrophysik, Karl-Schwarzschild-Strasse 1,
D-85741 Garching, Germany

This manuscript reviews recent progress in our understanding of the nucleosynthesis of medium and heavy elements in supernovae. Recent hydrodynamical models of core-collapse supernovae show that a large amount of proton-rich matter is ejected. This matter constitutes the site of the νp-process where antineutrino absorption reactions catalyze the nucleosynthesis of nuclei with $A > 64$. Supernovae are also associated with the r-process responsible for the synthesis of the heaviest elements in nature. Fission during the r-process can play a major role in determining the final abundance pattern and in explaining the almost universal features seen in metal-poor r-process-rich stars.

1. Introduction

In recent years, observations of metal-poor stars have contributed to increase our understanding of the nucleosynthesis of medium and heavy nuclei and its evolution during the history of the galaxy. Metal-poor stars with large enhancements of r-process elements (the abundance of Eu is typically considered to represent the presence of heavy r-process nuclei) with respect to iron show a variation of two to three orders of magnitude in the absolute amount of r-process elements present for stars of similar metallicities[1]. However the relative abundance of elements heavier that $Z > 56$ (but not including the radioactive actinides) shows a striking consistency with the observed solar abundances of these elements[1]. This consistency does not extend to elements lighter than $Z = 56$ where some variations are observed. In most of the cases elements lighter than $Z < 56$ are underabundant when compared with a scaled solar abundance distribution that matches the observed heavy element abundances[1]. However, recent observations of the metal-poor star HD 221170[2] show that in some cases the agreement between the scaled solar r-process abundance pattern and the observed abundances of elements can be extended to elements heavier than $Z > 37$. All these observations indicate that the astrophysical sites for the synthesis of light and heavy neutron capture elements are different[3,4] suggesting two disting r-processes. Possible sites are supernovae and neutron-star mergers. The exact site and operation for both types of r-process is not known, however, there are clear indications that while the process responsible for the production of heavy elements is universal[5] the production of lighter elements (in particular Sr, Y and Zr) has a much more complex Galactic history[6].

Even if the astrophysical site of the r-process(es) is (are) unknown, it is clear that the process is of primary nature. This means that the site has to produce both the neutrons and seeds necessary for the occurrence of a phase with fast neutron captures that characterizes the r-process[7]. Moreover, in order to explain the observed abundances of U and Th the neutron-to-seed ratio needs to be larger than ~ 100. Under these conditions fission of r-process nuclei beyond U and Th can play a mallor role in explaining the universality of the heavy r-process pattern in metal-poor stars. This issue will be discussed in section 3.

In section 2 we present a new nucleosynthesis process that we denote the νp-process which occurs in proton-rich matter ejected under explosive conditions and in the presence of strong neutrino fluxes. This process seems necessary to explain the observed abundances of light p-nuclei, including ^{92,94}Mo and ^{96,98}Ru.

2. Nucleosynthesis in proton-rich supernova ejecta

Recent hydrodynamical studies of core-collapse supernovae have shown that the bulk of neutrino-heated ejecta during the early phases (first second) of the supernova explosion is proton-rich[8,9,10]. Nucleosynthesis studies in this environment have shown that these ejecta could be responsible for the solar abundances of elements like ^{45}Sc, ^{49}Ti and ^{64}Zn [11,12]. Once reactions involving alpha particles freeze out, the composition in these ejecta is mainly given by $N = Z$ alpha nuclei and free protons. Proton captures on this nuclei cannot proceed beyond ^{64}Ge due to the low proton separation energy of ^{65}As and the fact that the beta-decay half-life of ^{64}Ge (64 s) is much longer than the typical expansion time scales (a few seconds). However, the proton densities and temperatures in these ejecta resemble those originally proposed for the p-process by B^{2}FH[13]. So it is interesting to ask under which conditions the nucleosynthesis flow can proceed beyond ^{64}Ge and contribute to the production of light p-nuclei like ^{92,94}Mo and ^{96,98}Ru that are systematically underproduced in other scenarios[14].

Two recent studies[15,16] have shown that the inclusion of neutrino interactions during the nucleosynthesis permits a new chain of nuclear reactions denoted νp-process in ref.[15]. In this process nuclei form at a typical distance of $\sim$ 1000 km from proto-neutron star where antineutrino absorption reactions proceed on a time scale of seconds that is much shorter than the typical beta decay half-lives of the most abundant nuclei present (eg. ^{56}Ni and ^{64}Ge). As protons are more abundant than heavy nuclei, antineutrino capture occurs predominantly on protons via $\bar{\nu}_e + p \rightarrow n + e^+$, causing a residual density of free neutrons of 10^{14}–10^{15} cm^{-3} for several seconds, when the temperatures are in the range 1–3 GK. These neutrons can easily be captured by neutron-deficient $N \sim Z$ nuclei (for example ^{64}Ge), which have large neutron capture cross sections. The amount of nuclei with $A > 64$ produced is then directly proportional to the number of antineutrinos captured. While proton capture, (p, γ), on ^{64}Ge takes too long, the (n, p) reaction dominates (with a lifetime of 0.25 s at a temperature of 2 GK), permitting the matter flow to continue to heavier nuclei than ^{64}Ge via subsequent proton captures and beta decays till the next alpha nucleus, ^{68}Se. Here again (n, p) reactions followed by proton captures and beta decays permit the flow to reach heavier alpha nuclei. This process can continue till proton capture reactions freeze out at temperatures around 1 GK. The νp-process is different to r-process nucleosynthesis in environtments with $Y_e < 0.5$, i.e. neutron-rich ejecta, where neutrino captures on neutrons provide protons that interact mainly with the existing neutrons, producing alpha-particles and light nuclei. Proton capture by heavy nuclei is suppressed because of the large Coulomb barriers[17,18]. Consequently, in r-process environments an enhanced formation of the heaviest nuclei does not take place when neutrino are present. In proton-rich ejecta, in contrast to expectation[12], antineutrino absorption produces neutrons that do not suffer from Coulomb barriers and are captured preferentially by heavy neutron-deficient nuclei.

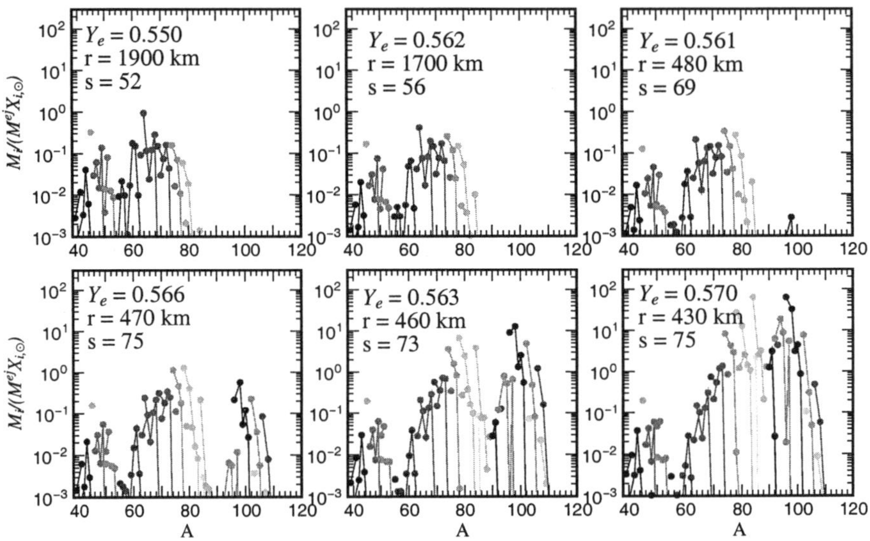

Figure 1. Production factors for six hydrodynamical trajectories corresponding to the early proton rich wind obtained in the explosion of a 15 $M_\odot$ star [8]. In each panel the radius, entropy and Y_e values of matter when the temperature reaches 3 GK are shown.

As discussed above the νp-process acts in the temperature range of 1–3 GK. The amount of heavy nuclei synthesized depends on the ratio of neutrons produced via antineutrino capture to the abundance of heavy nuclei (this is similar to the neutron-to-seed ratio in the r-process, see also discussion in[16]). This ratio is sensitive to the antineutrino fluence and to the proton to seed ratio. The first depends mainly on the expansion time scale of matter and its hydrodynamical evolution. The second is very sensitive to the proton richness of the material and its entropy. Figure 1 shows the nucleosynthesis resulting from several trajectories corresponding to the early proton-rich wind from the protoneutron star resulting of the explosion of a 15 $M_\odot$ star[8]. (These trajectories have also been studied in reference[16].) No production of nuclei above $A = 64$ is obtained if antineutrino absorption reactions are neglected. Once they are included production of elements above $A = 64$ takes place via the chain of reactions discussed in the previous paragraph. This allows to extend the nucleosynthesis beyond Zn producing elements like Ge whose abundance is roughly proportional to the iron abundance at low metallicities[1]. The production of light p-nuclei like ^{84}Sr, ^{94}Mo and ^{96,98}Ru is also clearly seen in figure 1. However, ^{92}Mo is still underproduced. This could be due to the limited knowledge of masses around ^{92}Pd[16]. The current mass systematics[19] predict a rather low proton separation energy for ^{91}Rh that inhibits the production of ^{92}Pd. Future experimental work in this region should clarify this issue. However, this could be a feature of the νp-process. In this case, it is interesting to notice that previous studies have shown that ^{92}Mo can be produced in slightly neutron-rich ejecta with $Y_e \approx 0.47$–0.49[17,20]. A recent study[21] has shown that a combination of proton-rich and slightly neutron-rich ejecta produces all light p-nuclei. Certainly, much work needs to be done in order to understand the transition from proton-rich to neutron-rich matter in consistent supernovae simulations and its dependence with stellar mass.

3. The role of fission in the r-process

The r-process is responsible for the synthesis of at least half of the elements heavier than Fe. It is associated with explosive scenarios where large neutrons densities are achieved allowing for the series of neutron captures and beta decays that constitutes the r-process[7,22]. The r-process requires the knowledge of masses and beta-decays for thousands of extremely neutron-rich nuclei reaching even the neutron-drip line. Moreover, in order to synthesize the heavy long-lived actinides, U and Th, large neutron to seed ratios are required ($\sim$ 100) allowing to reach nuclei that decay by fission. Fission can be induced by different processes: spontaneous fission, neutron induced fission, beta-delayed fission and, if the r-process occurs under strong neutrino fluxes, neutrino-induced fission. The role of fission in the r-process has been the subject of many studies in the past (see ref.[23] and references therein), however, often only a subset of fission-inducing reactions was considered and a rather simplistic description of fission yields was used. It should be emphasized that, if fission really plays a role in determining the final abundances of the r-process, one needs not only fission rates but equally important are realistic fission yields as they determine the final abundances. Our goal has been to improve this situation by putting together a full set of fission rates including all possible fission reactions listed above. We use the Thomas-Fermi fission barriers of reference[24] which accurately reproduce the isospin dependence of saddle-point masses[25]. The neutron-induced fission rates are from reference[23]. Beta-delayed fission rates are determined based on the FRDM beta-decay rates[26] using an approximate strength distribution for each decay build on the neutron-emission probabilities[a]. The spontaneous fission rates are determined by a regression fit of experimental data[27] to the Thomas-Fermi fission barriers. For each fissioning nucleus the fission yields are determined using the statistical code ABLA[28,29]. The fission yields change from nucleus to nucleus and in a given nucleus depend on the excitation energy at which fission is induced.

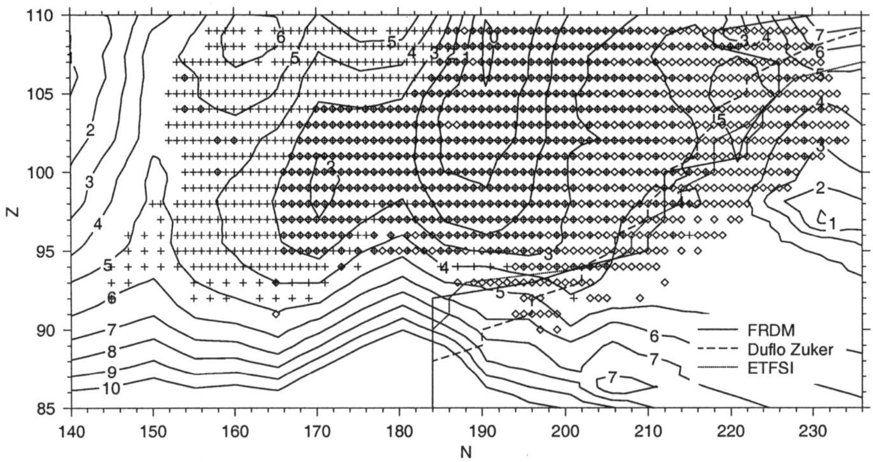

Figure 2. Region of the nuclear chart where fission takes place during the r-process. The contour lines represent the Thomas-Fermi fission barrier heights in MeV. Crosses show the nuclei for which neutron-induced fission dominates over (n, γ). Diamonds show the nuclei for which the spontaneous fission or beta-delayed fission operates in a time scale smaller than 1 second. The lines show the location for which negative neutron separation energies are found in different mass models (FRDM[30], ETFSI[31] and Duflo-Zuker[32]).

Figure 2 shows the region where fission takes place during the r-process. When the r-process reaches nuclei with $Z \sim 85$–90 matter accumulates at the magic neutron number $N = 184$ that plays a similar role as the standard waiting points at $N = 82$ and 126. Nuclei in this mass range have large fission barriers so that fission is only possible once matter moves beyond $N = 184$. The amount of matter that is able to proceed beyond this point depends of the magnitude of the $N = 184$ shell gap. The Duflo-Zuker[32] mass model shows the weakest shell gap, while masses based on the ETFSI model[31] show the stronger shell gap; the FRDM model[30] is somewhat in between. Once matter has passed $N = 184$, neutron-induced fission takes place in the region $Z \sim 90$–95 and $N \sim 190$. Once fission occurs, the main consequence is that neutrons are mainly captured by fis-

[a]http://t16web.lanl.gov/Moller/publications/tpnff.dat

sioning nuclei that have a larger net capture rate (difference between the capture and its inverse process). Once a neutron induces a fission the fissioning nucleus emits around 2–4 neutrons during the fission process. But a larger amount of neutrons is produced by the decay of the fission products which have a Z/A ratio similar to the fissioning nucleus so that they are located closer to the neutron-drip line than the r-process path. Thus, the fission products will decay either by photodissociation, (γ, n), or beta decays (mainly by beta-delayed neutron emission) to the r-process path, emitting of order 8 neutrons per fragment. This implies that each neutron-induced fission event produces around 20 neutrons.

Once neutrons are exhausted the matter accumulated at $N = 184$ beta-decays producing neutrons by beta-delayed neutron emission. These neutrons induce new fissions in the region $Z \sim 95$, $N \sim 175$ that is fed by beta-decays and produce more neutrons self-enhancing the neutron-induced fission rate by a mechanism similar to a chain reaction. The net result is that neutron-induced fission dominates over beta-delayed or spontaneous fission as it can operate in time scales of less than a ms for neutron densities above 10^{18} cm^{-3}.

The above qualitative arguments which show the role of fission during the r-process are independent of the fission barriers used. To get a more quantitative understanding, we have carried out fully dynamical calculations that resemble the conditions expected in the high-entropy bubble resulting in a core-collapse supernova explosion. Early calculations[34] failed to produce the large entropies required for a successful r-process[35]. However, recent calculations indicate that the high entropies required by the r-process can be attained[36]. In our calculations, we assume an adiabatic expansion of the matter, as described in reference[37], but using a realistic equation of state[38]. We adjust the entropy to produce large enough neutron-to-seed ratios to study the effect of fission. We notice that the neutron-to-seed ratio does not only depends on entropy, but also on neutron-richness and expansion time scale[35].

Figure 3 shows the results of our calculations for three different mass models. While the FRDM and Duflo-Zuker mass models show a similar trend with increasing neutron-to-seed ratio, the ETFSI-Q mass model is clearly different. This difference is due to the fact that the ETFSI-Q mass has a quenched shell gap for $N = 82$ and $N = 126$, while the other two mass models show strong shell gaps even close to the drip line. In the ETFSI-Q mass model the $N = 82$ waiting point is practically absent for the conditions of figure 3. This allows all matter to pass through $N = 82$, incorporating most neutrons in heavy nuclei and leaving a few free neutrons to induce fission events. In the other two models, a smaller amount of matter passes the $N = 82$ and $N = 126$ waiting points. Once this matter reaches the fissioning region a large abundance of neutrons is still present

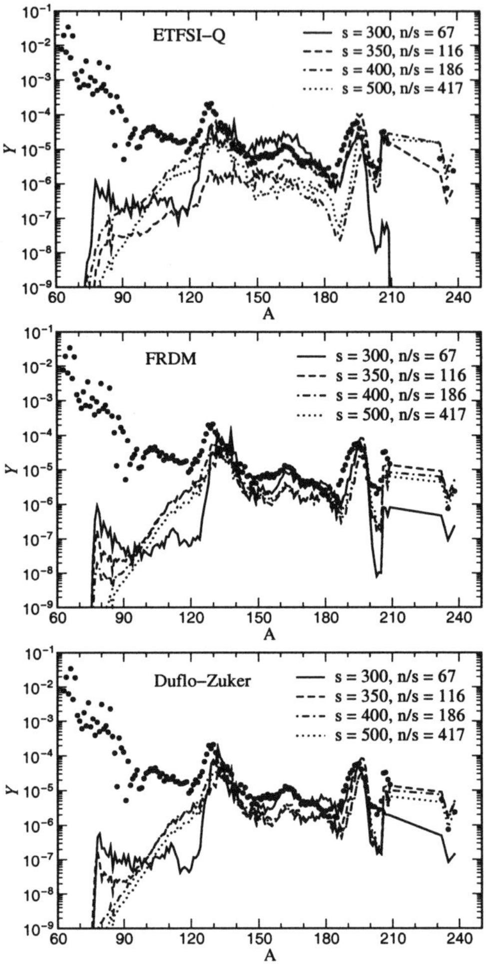

Figure 3. Final r-process abundances obtained in several adiabatic expansions using different mass models (FRDM[30], ETFSI-Q[33] and Duflo-Zuker[32]). All the calculation are done for a constant expansion velocity of 4500 km (corresponding to a dynamical time scale of 50 ms). The product ρr^3 is keep constant during the expansion and the temperature is determined from the equation of state under the condition of constant entropy. The curves are labeled according to the entropy and neutron to seed ratio (n/s) resulting after the alpha-rich freeze-out.

that creates new neutrons by fission allowing the r-process to last for a longer time and produce a larger fraction of fission fragments. This explains why the FRDM and Duflo-Zuker mass models produce larger amounts of matter in the range $A = 130$–190, and implies that the shell structure at $N = 82$ is essential for determining the role of fission in the r-process. Calculations with mass models with strong shell gaps yield final abundances that are practically independent of the conditions once the neutron-to-seed ratio is large enough. This seems to be consistent with metal-poor star observations that show a universal abundance distribution of elements heavier that $Z = 56^1$.

Our calculations also show that neutron-induced fission is the major fission process. For example for the calculations with neutron-to-seed ratio 186, the percentage of final abundance that has undergone neutron-induced fission is 36%, beta-delayed fission 3% and neutrino-induced fission 0.3%. With increasing neutron-to-seed ratio all the percentages increase but the relative proportions remain practically constant.

4. Conclusions

The study of the nucleosynthesis processes responsible for the production of medium and intermediate elements and their relationship to supernovae constitutes a challenge to astronomers, astrophysicists and nuclear physicists. Our current understanding is driven by high-resolution spectroscopic observations of metal-poor stars that aim to probe individual nucleosynthesis events. At the same time progress in the modeling of core-collapse supernovae has improved our knowledge of explosive nucleosynthesis in supernovae. In particular the presence of proton-rich ejecta has open the way to find a solution to the long-standing problem of the origin of light p-nuclei. Further progress will come from advances in the modeling of the supernovae explosion mechanism and from improved knowledge of the properties of the involved nuclei to be studied at future radioactive-ion beam facilities. These facilities will also open the door to the study many of the nuclei involved in r-process nucleosynthesis, in particular the nuclei located near the $N = 82$ waiting point that are important in determining the role of fission in the r-process. They will also provide valuable data needed to constrain theoretical models to allow for more reliable extrapolations to the region of the nuclear chart where fission takes place during r-process nucleosynthesis.

References

1. J. J. Cowan and C. Sneden, *Nature* **440**, 1151 (2006).
2. I. I. Ivans, J. Simmerer, C. Sneden, J. E. Lawler, J. J. Cowan, R. Gallino and S. Bisterzo, *Astrophys. J.* **645**, 613 (2006).
3. G. J. Wasserburg, M. Busso and R. Gallino, *Astrophys. J.* **466**, p. L109(August 1996).
4. Y.-Z. Qian and G. J. Wasserburg, *Astrophys. J.* **559**, 925 (2001).
5. S. Wanajo and Y. Ishimaru, *Nucl. Phys. A* (2006), in press.

6. C. Travaglio, R. Gallino, E. Arnone, J. Cowan, F. Jordan and C. Sneden, *Astrophys. J.* **601**, 864 (2004).
7. J. J. Cowan, F.-K. Thielemann and J. W. Truran, *Phys. Repts.* **208**, 267 (1991).
8. R. Buras, M. Rampp, H.-T. Janka and K. Kifonidis, *Astron. & Astrophys.* **447**, 1049 (2006).
9. M. Liebendörfer, A. Mezzacappa, F.-K. Thielemann, O. E. Bronson Messer, W. Raphael Hix and S. W. Bruenn, *Phys. Rev. D* **63**, 103004 (2001).
10. T. A. Thompson, E. Quataert and A. Burrows, *Astrophys. J.* **620**, 861 (2005).
11. C. Fröhlich, P. Hauser, M. Liebendörfer, G. Martínez-Pinedo, F.-K. Thielemann, E. Bravo, N. T. Zinner, W. R. Hix, K. Langanke, A. Mezzacappa and K. Nomoto, *Astrophys. J.* **637**, 415 (2006).
12. J. Pruet, S. E. Woosley, R. Buras, H.-T. Janka and R. D. Hoffman, *Astrophys. J.* **623**, 325 (2005).
13. E. M. Burbidge, G. R. Burbidge, W. A. Fowler and F. Hoyle, *Rev. Mod. Phys.* **29**, 547 (1957).
14. M. Arnould and S. Goriely, *Phys. Rep.* **384**, 1 (2003).
15. C. Fröhlich, G. Martínez-Pinedo, M. Liebendörfer, F.-K. Thielemann, E. Bravo, W. R. Hix, K. Langanke and N. T. Zinner, *Phys. Rev. Lett.* **96**, 142502 (2006).
16. J. Pruet, R. D. Hoffman, S. E. Woosley, H.-T. Janka and R. Buras, *Astrophys. J.* **644**, 1028 (2006).
17. G. M. Fuller and B. S. Meyer, *Astrophys. J.* **453**, 792 (1995).
18. B. S. Meyer, G. C. McLaughlin and G. M. Fuller, *Phys. Rev. C* **58**, 3696 (1998).
19. G. Audi, A. H. Wapstra and C. Thibault, *Nucl. Phys. A* **729**, 337 (2003).
20. R. D. Hoffman, S. E. Woosley, G. M. Fuller and B. S. Meyer, *Astrophys. J.* **460**, 478 (1996).
21. S. Wanajo, *Astrophys. J.* (2006), in press.
22. J. J. Cowan and F.-K. Thielemann, *Physics Today* , 47(October 2004).
23. I. V. Panov, E. Kolbe, B. Pfeiffer, T. Rauscher, K.-L. Kratz and F.-K. Thielemann, *Nucl. Phys. A* **747**, 633 (2005).
24. W. D. Myers and W. J. Świątecki, *Phys. Rev. C* **60**, 014606 (1999).
25. A. Kelić and K.-H. Schmidt, *Phys. Lett. B* **634**, 362 (2005).
26. P. Möller, B. Pfeiffer and K.-L. Kratz, *Phys. Rev. C* **67**, 055802 (2003).
27. T. Kodama and K. Takahashi, *Nucl. Phys. A* **239**, 489 (1975).
28. J.-J. Gaimard and K.-H. Schmidt, *Nucl. Phys. A* **531**, p. 709 (1991).
29. J. Benlliure, A. Grewe, M. de Jong, K.-H. Schmidt and S. Zhdanov, *Nucl. Phys. A* **628**, p. 458 (1998).
30. P. Möller, J. R. Nix and K.-L. Kratz, *At. Data. Nucl. Data Tables* **66**, p. 131 (1997).
31. Y. Aboussir, J. M. Pearson, A. K. Duttab and F. Tondeur, *Nucl. Phys. A* **549**, 155 (1992).
32. J. Duflo and A. P. Zuker, *Phys. Rev. C* **52**, R23 (1995).
33. J. M. Pearson, R. C. Nayak and S. Goriely, *Phys. Lett. B* **387**, 455 (1996).
34. J. Witti, H.-T. Janka and K. Takahashi, *Astron. & Astrophys.* **286**, 857 (1994).
35. R. D. Hoffman, S. E. Woosley and Y.-Z. Qian, *Astrophys. J.* **482**, p. 951 (1997).
36. A. Burrows, E. Livne, L. Dessart, C. D. Ott and J. Murphy, *Astrophys. J.* **640**, 878 (2006).
37. C. Freiburghaus, J.-F. Rembges, T. Rauscher, E. Kolbe, F.-K. Thielemann, K.-L. Kratz,

B. Pfeiffer and J. J. Cowan, *Astrophys. J.* **516**, 381(May 1999).
38. F. X. Timmes and D. Arnett, *Astrophys. J. Suppl.* **125**, 277 (1999).

THE IMPORTANCE OF THE RP-PROCESS IN THERMONUCLEAR BURNING ON ACCRETING NEUTRON STARS

ANDREW CUMMING

Physics Department, McGill University, 3600 rue University, Montreal, QC, H3A 2T8, Canada; cumming@physics.mcgill.ca

I summarize recent work on two aspects of nuclear burning on accreting neutron stars that are particularly sensitive to the details of the rp-process: the lightcurves of Type I X-ray bursts which show long tails powered by rp-process burning, and "superbursts", which are long duration and energetic bursts powered by ignition of carbon in the rp-process ashes. Models of Type I X-ray bursts show remarkable agreement with observed lightcurves from GS 1826-24, including variations in the lightcurve with accretion rate. Models of superburst lightcurves suggest that the carbon mass fraction is 10–20% at ignition. Production of such large carbon fractions at the observed accretion rates is not understood.

1. Introduction

Nuclear burning on accreting neutron stars in low mass X-ray binaries (LMXBs) is often thermally unstable, rendering it visible to the observer as short bursts of X-rays known as Type I X-ray bursts[1]. As the accreted fuel accumulates on the star (typically at rates 10^{-11}–10^{-8} $M_{\odot}$ yr^{-1}), it is usually hot enough that the hydrogen burns stably via the beta-limited hot CNO cycle[2]. However, once the base of the accumulating layer reaches densities of $\approx 10^5$ g cm^{-3} and temperatures of $\approx 2 \times 10^8$ K, helium ignites by the triple alpha reaction which is very temperature sensitive. Temperature fluctuations then drive a thermal instability that rapidly burns the accreted fuel[3,4,5]. The thermonuclear flash is observed as a brief (10–100 s) burst of X-rays, with a rapid rise followed by a slower, often exponential-like, decay as the layer cools.

The burning of hydrogen and helium in these events is only the first step in a sequence of nuclear processes that transform the accreted matter as it is pushed deeper into the star by continued accretion. From deeper carbon burning[6,7], to electron captures and pycnonuclear reactions in the crust[8,9], to the equation of state of the dense core[10], nuclear physics input is crucial every step of the way. Studies of nuclear burning in these systems also address a wide range of other astrophysics questions, including the origin and evolution of neutron star spin and magnetic fields[11], evolution of binary systems, particularly short orbital period ultracompact binaries[12,13], and the physics of how the burning propagates[14].

The high temperatures reached during the flash ($\gtrsim 2 \times 10^9$ K) lead to breakout from the CNO cycle, and subsequent burning of the hydrogen and helium by αp- and rp-processes[15,16]. Modeling the rp-process requires accurate nuclear masses, proton capture and beta decay rates for heavy proton-rich nuclei well beyond the iron group[17,18]. Much of this data is not or is poorly known experimentally. In terms of the observable properties of the nuclear burning, three characteristics of the rp-process are important: its timescale, its energetics, and the composition of the ashes. Nuclear burning on accreting neutron stars and the role of nuclear physics has been recently reviewed by Schatz and Rehm[17]. Here, I focus on two aspects of nuclear burning which are particularly sensitive to the details of the rp-process: the lightcurves of Type I X-ray bursts which show long tails powered by rp-process burning, and the rare, energetic "superbursts", which we believe are powered by ignition of carbon in the rp-process ashes.

2. Observing the rp-process in burst lightcurves

During the X-ray burst, the cooling time of the burning layer is typically a few seconds, so that prolonged energy release by the rp-process has a significant effect on the burst lightcurve, giving rise to a long tail lasting for a minute or longer. The sensitivity of the burst lightcurve shape to the nuclear physics input was first studied in one-zone models[19,20,21]. By integrating over the vertical structure of the layer, the calculation is greatly simplified since only a single temperature, pressure and composition has to be followed. In these calculations, different choices for the nuclear physics input gave dramatic variations in the duration and shape of the burst lightcurve. While these results suggested that burst lightcurves would provide an interesting test of our understanding of the rp-process, one open question was whether these variations would persist when the detailed vertical structure of the layer, including mixing due to convection, was included.

In fact, "multi-zone" calculations are now being carried out and show that the lightcurves are affected by changes in the nuclear physics input. These calculations fully resolve the burning layer in spherical symmetry, and follow the nucleosynthesis at each depth through a series of successive bursts[22,23,24]. Woosley *et al.* (ref. 22) carried out a first study of the effects of nuclear physics uncertainties by varying all of the beta decay timescales along the rp-process path by a constant factor, finding considerable variation in the predicted lightcurve. Fisker *et al.* have shown[23] that the double peaked nature of some observed X-ray bursts may be due to a temporary delay in the rp-process at the waiting point associated with ^{32}S. In a

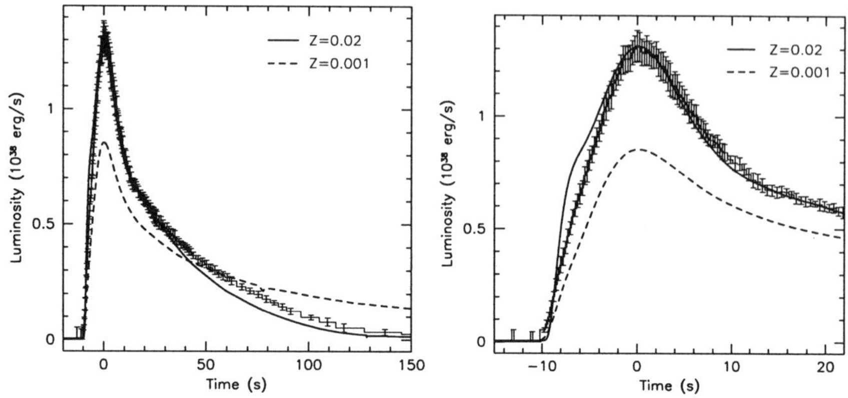

Figure 1. From Heger *et al.* (ref. 28). Comparison between the average burst lightcurve observed from GS 1826-24 (data points with error bars), and the theoretical burst models (solid line for solar metallicity, dashed line for $1/20$ solar metallicity). The distance to the source has been chosen as 6.4 kpc, which brings the peak luminosities of the observed and predicted lightcurves into agreement.

different calculation[24], they find that variations in the $^{15}\mathrm{O}(\alpha, \gamma)$ breakout reaction rate within current uncertainties lead to qualitatively different behavior, including complete suppression of the thermal instability for the lowest breakout rates.

The best source for comparison with these calculations is the "clocked"[25] or "textbook"[26] burster GS 1826-24. Type I X-ray bursts from this source repeat extremely regularly (on timescales between 3 to 6 hours depending on the accretion rate), with lightcurves that are remarkably similar from burst to burst. This behaviour nicely matches the simple prediction of theory: a limit cycle involving repeated accumulation and burning of fuel. The steady burst rate probably indicates a very steady supply of fuel to the star. At the accretion rate $\dot{M} \approx 10^{17}$ g s^{-1} inferred from the luminosity of the source ($L_X \approx 10^{37}$ erg s^{-1}), a mass $\Delta M \approx 10^{21}$ g accumulates in a few hours. This matches the ignition mass predicted by burst models[27,26]. In the few hours between bursts, the hot CNO cycle does not have sufficient time to burn all of the hydrogen away, so that helium ignition occurs in a hydrogen-rich environment and the rp-process ensues.

Figure 1 shows a comparison[28] of the GS 1826-24 burst lightcurve observed[29] with the *Rossi X-ray Timing Explorer (RXTE)* and the simulations of Woosley *et al.* (ref. 22). The agreement with the solar metallicity model is remarkable. Note that the distance to the source has been chosen within the uncertainties so that the peak luminosities of the observed and solar metallicity model lightcurves agree (the required distance is 6.4 kpc). However, there is no distance which gives good agreement with the low metallicity model: the shape of the lightcurve in

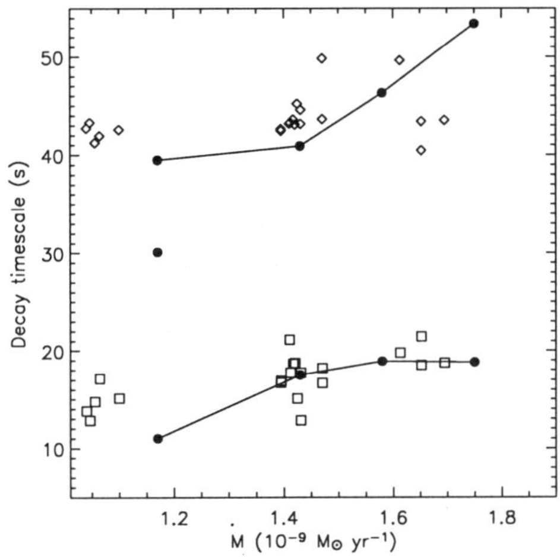

Figure 2. From Heger *et al.* (ref. 28). Fitted exponential decay times for observed bursts (open symbols) and theoretical lightcurves (solid symbols). In each case, the shorter timescale refers to the decay immediately after the peak; the longer timescale refers to the burst tail.

the low metallicity model is clearly different than observed. The difference in shape between the low metallicity and solar metallicity models is a consequence of hydrogen burning between bursts by the hot CNO cycle. Burst simulations show that helium burning powers the initial stage of the burst, which then gives way to slow rp-process burning in the tail[30,22]. In the low metallicity model, very little CNO burning occurs between bursts, leading to a larger hydrogen fraction (and smaller helium fraction) at ignition than in the solar metallicity model. The larger proton to seed ratio then gives a more extended rp-process and a longer tail, while the smaller helium fraction leads to a less energetic initial stage of the burst.

The theoretical models with solar metallicity also reproduce the changes in the lightcurve observed as the accretion rate varies. Over a period of several years the recurrence time decreased as the source flux F_X (and presumably accretion rate since we expect $F_X \propto \dot{M}$) increased[29]. The observed relation is close to $\Delta t \propto \dot{M}^{-1}$, or constant ignition mass, in agreement with theoretical predictions[27]. As the recurrence time dropped, the shape of the burst lightcurves changed slightly. In particular, the initial part of the burst became less luminous and decayed less quickly. Galloway *et al.* (ref. 29) suggested that this change is due to a changing composition at the time of ignition. With a shorter recurrence time, less hot CNO burning occurs between bursts, giving a smaller helium fraction at ignition and a less luminous and energetic initial stage of the burst. Figure 2 quantifies this effect, by comparing the fitted exponential decay times of the observations with theoretical models; they show good agreement. The idea that hot CNO burning is occurring between bursts is also suggested by variations in the burst energetics, which show that the bursts become more energetic when the recurrence time drops (more hydrogen present at ignition). Galloway *et al.* (ref. 29) used these variations to argue for solar metallicity in this source, a conclusion that is supported by the lightcurve comparisons.

3. Superbursts and the rp-process ashes

Superbursts are energetic Type I X-ray bursts which have been discovered by long term monitoring of X-ray bursters with *RXTE* and *BeppoSAX*. They have durations of several hours, energies $\approx 10^{42}$ ergs, and recurrence times $\approx$ 1–2 yrs. Each of these numbers is 100–1000 times larger than usual Type I X-ray bursts. Currently, 13 superbursts have been seen from 9 sources[31,32]. One of the most fascinating features of superbursts is that they "interact" with normal Type I X-ray bursts. For approximately a month following a superburst, Type I X-ray bursts are "quenched", presumably by the large outwards flux as the ocean cools[6,33]. Also, precursor bursts with similar properties to normal Type I X-ray bursts are observed immediately before superbursts, likely due to ignition of an overlying helium layer as the superburst ignites. Together with their rarity, large energies and durations, these interactions point to superbursts arising from a deeper fuel layer.

Superbursts are thought to be thermonuclear flashes driven by carbon burning in the deep ocean of the neutron star[6,7]. The idea that carbon burning could be unstable on accreting neutron stars is an old one, for example once proposed as a gamma-ray burst model in the 1970's[34]. These models, which assumed ignition of a pure carbon layer, predicted an energy release 100 times larger than observed in superbursts. However, the amount of carbon expected in the rp-process ashes is small, and in addition, the heavy elements which make up most of the ashes act to make the fuel layer opaque, increasing the temperature and leading to earlier ignition. For an assumed carbon mass fraction $X_C = 10\%$ and heavy elements near the rp-process endpoint[20], Cumming & Bildsten (ref. 6) found ignition depths of $\approx 10^{12}\ \mathrm{g\ cm^{-2}}$ giving energies $\approx 10^{42}$ erg, in agreement with observations. This amount of mass is accreted in 1–3 years at 0.1–0.3 $\dot{M}_{\rm Edd}$, consistent with observational limits on the recurrence times ($\dot{M}_{\rm Edd} \approx 10^{18}\ \mathrm{g\ s^{-1}}$ is the Eddington accretion rate).

There is evidence for carbon mass fractions of 10–20% from the superburst lightcurves. Cumming and Macbeth (ref. 33) modeled the lightcurves by assuming instantaneous burning of the fuel, and then following the thermal evolution of the layer as it cooled. The luminosity during the early part of the lightcurve depends directly on the energy released in the layer. Fits to observed superburst lightcurves imply that an energy release of $\approx 2 \times 10^{17}$ erg g^{-1} occurs[35], which corresponds to $X_C \approx 20\%$ for an energy release of 1 MeV per nucleon from carbon burning to iron group. A smaller carbon fraction can be accommodated if the heavy elements in the rp-process ashes undergo photodisintegration during the flash, which can double the total energy release[36], giving $X_C \approx 10\%$. The inferred energy release is quite well constrained; if $E_{17} \approx 1$, the energy released during the first few hours would be substantially less than observed, while if $E_{17} \gtrsim 3$, the luminosity would exceed the Eddington luminosity for several minutes or longer, whereas there is no evidence for long durations of photospheric radius expansion (except for the superburst from the pure He accretor 4U 1820-30, see ref. 7), and the peak superburst luminosity is often small compared to the maximum luminosity of normal Type I X-ray bursts from the same source.

Carbon mass fractions of 10–20% are difficult to make at the accretion rates $\dot{M} = 0.1$–$0.3\ \dot{M}_{\rm Edd}$ at which superbursts are observed. The reason is that hydrogen and helium burning is expected to be unstable at these accretion rates (giving rise to Type I X-ray bursts). In unstable burning, the high temperatures result in rapid consumption of the helium before the hydrogen runs out. Therefore any carbon produced by the triple alpha reaction is immediately "poisoned" by proton capture. Both one-zone[37] and multizone calculations[22] find $X_C < 1\%$ from unstable burning. Such low carbon fractions do not lead to a thermal instability; instead the carbon quietly burns away as the ashes accumulate[6].

Stable hydrogen/helium burning is much better at producing carbon[38]. Since the burning occurs at lower temperature, the hydrogen runs out before the helium burns completely away[39], allowing carbon to survive. In fact, there is observational evidence that stable hydrogen/helium burning occurs in superburst sources. In 't Zand *et al.* (ref. 40) found that superburst sources have large values of α, the ratio of persistent fluence between bursts (which measures the gravitational energy release of infalling matter) to the burst fluence (which measures the nuclear energy release in bursts). Superburst sources have $\alpha \sim 1000$, implying that only $\sim 10\%$ of the accreted fuel burns in bursts.

Calculations of stable burning at these accretion rates show that enough carbon is produced to explain the superbursts[38]. However, the situation is uncomfortable because the mechanism underlying stable burning at accretion rates $< \dot{M}_{\rm Edd}$ is not understood[26]. One possibility is that the rate of breakout reactions is less than currently thought[41], in which case accretion of material with metallicity greater than solar could lead to sufficient carbon being produced[42]. Another alternative is that the accreted material covers only part of the neutron star surface for global accretion rates $\gtrsim 0.1\ \dot{M}_{\rm Edd}$, increasing the local accretion rate and stabilizing the burning[43]. In that case, accretion of helium-rich material is likely required, since steady burning with solar composition at $\dot{M} \approx \dot{M}_{\rm Edd}$ gives a low carbon fraction $\approx 5\%$. Other mechanisms to stabilize burning have been proposed, such as rotational mixing[44], but the effect on carbon production has not been explored.

Finally, there has been a lot of recent interest in the possibility of using superbursts to probe the neutron star interior properties. Brown (ref. 45) first pointed out that the carbon ignition depth is very sensitive to the thermal properties of the crust and core because the accumulating carbon and heavy element layer is heated by the flux emerging from the crust. Different calculations[45,46,35] agree that reproducing the observed superburst ignition depths requires a large outwards heat flux (roughly 0.2–0.3 MeV per nucleon at $0.3\ \dot{M}_{\rm Edd}$). Achieving this heat flux is, however, currently a problem for two reasons. First, if Cooper pair neutrino emission is active in the crust, it limits the crust temperature to $\approx 4 \times 10^8$ K, too cold[35] for carbon ignition at 10^{12} g cm^{-2}. Second, superbursts are observed from transiently accreting systems, for which the outwards heat flux can be measured directly once accretion switches off at the end of the accretion outburst. In KS 1731-260, the measured quiescent flux[47] is more than an order of magnitude too low to achieve superburst ignition. Interestingly, recent work suggests that the Cooper pair emission rate may be overestimated[48], which would solve the first problem. However, the second problem remains; a consistent solution to explain superbursts and the quiescent luninosity of transients has not yet been found.

4. Conclusions

The long tails of Type I X-ray bursts from GS 1826-24 are thought to be powered directly by rp-process burning. The observed lightcurves agree remarkably well with multizone models. The good agreement is perhaps surprising given the large variations in lightcurve shape on varying the nuclear physics input. Further work is required to determine what we can learn from this comparison. The largest discrepancy between the observed and predicted lightcurves occurs in the burst rise. One possibility is that the treatment of nuclear physics or heat transport in the model at early times needs to be revised; alternatively, the profile of the rise could be set by spreading of the burning across the stellar surface.

Superbursts are a direct probe of the ashes of the rp-process. The timescale for the rp-process determines the point at which hydrogen runs out and therefore how much carbon is produced[38]. Observations of superbursts and modeling of their lightcurves[35] suggest that the carbon fraction is 10–20%. These large carbon fractions are a puzzle, since theory predicts that hydrogen and helium should burn unstably at the accretion rates of superburst sources, giving carbon fractions of $< 1\%$. Stable burning is likely required, and indeed observations of superburst sources show that they burn only $\approx 10\%$ of the accreted fuel in bursts. The problem of how to explain stable burning for accretion rates $\gtrsim 0.1\ \dot{M}_{\rm Edd}$ has been raised before in the context of normal Type I X-ray bursts[5,43]. Superbursts give new impetus to work on this question.

I have focused on these two aspects of nuclear burning on accreting neutron stars because they are particularly sensitive to the rp-process. Another example which I can only briefly mention here is the mHz quasi-periodic oscillations observed in the persistent X-ray flux from three accreting neutron stars[49], which have recently been identified with marginally stable nuclear burning on the neutron star surface[43]. This gives an opportunity to probe rp-process burning at the transition between unstable and stable burning, and therefore in different, lower temperature, conditions than during Type I X-ray bursts. Understanding all of these phenomena requires detailed nuclear physics input, and also at the same time gives an opportunity to test our understanding of the nuclear processes at work on accreting neutron stars.

Thanks to my collaborators in the work described here: D. Galloway, A. Heger, and S. Woosley on GS 1826-24, and J. in 't Zand, J. Macbeth, and D. Page on superburst models. I am grateful for support as a Alfred P. Sloan Research Fellow, and from the National Sciences and Engineering Research Council (NSERC), Le Fonds Québécois de la Recherche sur la Nature et les Technologies (FQRNT), and the Canadian Institute for Advanced Research (CIAR).

References

1. T. E. Strohmayer and L. Bildsten, in *Compact Stellar X-ray Sources* (Cambridge University Press, 2006)
2. F. Hoyle and W. A. Fowler, Nature of Strong Radio Sources, in *Quasi-Stellar Sources and Gravitational Collapse*, eds. I. Robinson, A. Schild and E. L. Schucking (1965).
3. C. J. Hansen and H. M. van Horn, *Astrophys. J.* **195**, 735 (1975).
4. M. Y. Fujimoto, T. Hanawa and S. Miyaji, *Astrophys. J.* **247**, 267 (1981).
5. L. Bildsten, Thermonuclear Burning on Rapidly Accreting Neutron Stars, in *NATO ASIC Proc. 515: The Many Faces of Neutron Stars.*, eds. R. Buccheri, J. van Paradijs and A. Alpar (1998).
6. A. Cumming and L. Bildsten, *Astrophys. J.* **559**, L127 (2001).
7. T. E. Strohmayer and E. F. Brown, *Astrophys. J.* **566**, 1045 (2002).
8. P. Haensel and J. L. Zdunik, *Astron. Astrophys.* **227**, 431 (1990).
9. P. Haensel and J. L. Zdunik, *Astron. Astrophys.* **404**, L33 (2003).
10. D. Page and S. Reddy, *Ann. Rev. Nucl. Part. Sci.* **56** (2006).

11. D. Chakrabarty, E. H. Morgan, M. P. Muno, D. K. Galloway, R. Wijnands, M. van der Klis and C. B. Markwardt, *Nature* **424**, 42 (2003).
12. A. Cumming, *Astrophys. J.* **595**, 1077 (2003).
13. J. J. M. in't Zand, A. Cumming, M. V. van der Sluys, F. Verbunt and O. R. Pols, *Astron. Astrophys.* **441**, 675 (2005).
14. A. Spitkovsky, Y. Levin and G. Ushomirsky, *Astrophys. J.* **566**, 1018 (2002).
15. R. K. Wallace and S. E. Woosley, *Astrophys. J. Suppl.* **45**, 389 (1981).
16. H. Schatz, A. Aprahamian, J. Goerres, M. Wiescher, T. Rauscher, J. F. Rembges, F.-K. Thielemann, B. Pfeiffer, P. Moeller, K.-L. Kratz, H. Herndl, B. A. Brown and H. Rebel, *Phys. Rep.* **294**, 167 (1998).
17. H. Schatz and K. E. Rehm, *astro-ph/0607624* (2006).
18. H. Schatz, *astro-ph/0607625* (2006).
19. B. A. Brown, R. R. Clement, H. Schatz, A. Volya and W. A. Richter, *Phys. Rev. C* **65**, 045802 (2002).
20. H. Schatz, A. Aprahamian, V. Barnard, L. Bildsten, A. Cumming, M. Ouellette, T. Rauscher, F.-K. Thielemann and M. Wiescher, *Physical Review Letters* **86**, 3471 (2001).
21. O. Koike, M. Hashimoto, K. Arai and S. Wanajo, *Astron. Astrophys.* **342**, 464 (1999).
22. S. E. Woosley, A. Heger, A. Cumming, R. D. Hoffman, J. Pruet, T. Rauscher, J. L. Fisker, H. Schatz, B. A. Brown and M. Wiescher, *Astrophys. J. Suppl.* **151**, 75 (2004).
23. J. L. Fisker, F.-K. Thielemann and M. Wiescher, *Astrophys. J.* **608**, L61 (2004).
24. J. L. Fisker, J. Gorres, M. Wiescher and B. Davids, *astro-ph/0410561* (2004).
25. M. Cocchi, A. Bazzano, L. Natalucci, P. Ubertini, J. Heise, E. Kuulkers and J. J. M. in't Zand, *Advances in Space Research* **28**, 375 (2001).
26. L. Bildsten, Theory and Observations of Type I X-Ray Bursts from Neutron Stars, in *Rossi2000: Astrophysics with the Rossi X-ray Timing Explorer. March 22-24, 2000 at NASA's Goddard Space Flight Center, Greenbelt, MD USA, meeting abstract*, ed. T. E. Strohmayer (2000).
27. A. Cumming and L. Bildsten, *Astrophys. J.* **544**, 453 (2000).
28. A. Heger, A. Cumming, D. K. Galloway and S. E. Woosley, *in preparation* (2006).
29. D. K. Galloway, A. Cumming, E. Kuulkers, L. Bildsten, D. Chakrabarty and R. E. Rothschild, *Astrophys. J.* **601**, 466 (2004).
30. T. Hanawa, D. Sugimoto and M.-A. Hashimoto, *Pub. Astron. Soc. Japan* **35**, 491 (1983).
31. E. Kuulkers, *Nuclear Physics B Proceedings Supplements* **132**, 466 (2004).
32. J. J. M. in't Zand, R. Cornelisse and A. Cumming, *Astron. Astrophys.* **426**, 257 (2004).
33. A. Cumming and J. Macbeth, *Astrophys. J.* **603**, L37 (2004).
34. S. E. Woosley and R. E. Taam, *Nature* **263**, 101 (1976).
35. A. Cumming, J. Macbeth, J. J. M. i. Zand and D. Page, *Astrophys. J.* **646**, 429 (2006).
36. H. Schatz, L. Bildsten and A. Cumming, *Astrophys. J.* **583**, L87 (2003).
37. O. Koike, M.-a. Hashimoto, R. Kuromizu and S.-i. Fujimoto, *Astrophys. J.* **603**, 242 (2004).
38. H. Schatz, L. Bildsten, A. Cumming and M. Ouellette, *Nuclear Physics A* **718**, 247 (2003).
39. H. Schatz, L. Bildsten, A. Cumming and M. Wiescher, *Astrophys. J.* **524**, 1014 (1999).
40. J. J. M. in't Zand, E. Kuulkers, F. Verbunt, J. Heise and R. Cornelisse, *Astron. Astro-*

phys. **411**, L487 (2003).
41. R. L. Cooper and R. Narayan, *astro-ph/0608068* (2006).
42. R. L. Cooper, B. Mukhopadhyay, D. Steeghs and R. Narayan, *Astrophys. J.* **642**, 443 (2006).
43. A. Heger, A. Cumming and S. E. Woosley, *astro-ph/0511292* (2005).
44. M. Y. Fujimoto, M. Sztajno, W. H. G. Lewin and J. van Paradijs, *Astrophys. J.* **319**, 902 (1987).
45. E. F. Brown, *Astrophys. J.* **614**, L57 (2004).
46. R. L. Cooper and R. Narayan, *Astrophys. J.* **629**, 422 (2005).
47. R. E. Rutledge, L. Bildsten, E. F. Brown, G. G. Pavlov, V. E. Zavlin and G. Ushomirsky, *Astrophys. J.* **580**, 413 (2002).
48. L. B. Leinson and A. Pérez, *Physics Letters B* **638**, 114 (2006).
49. M. Revnivtsev, E. Churazov, M. Gilfanov and R. Sunyaev, *Astron. Astrophys.* **372**, 138 (2001).

FREE-NUCLEON/ALPHA-PARTICLE DISEQUILIBRIUM AND R-PROCESS NUCLEOSYNTHESIS

B. S. MEYER* and C. WANG

Department of Physics and Astronomy, Clemson University,
Clemson, SC 29634-0978, USA
**E-mail: mbradle@clemson.edu*
`http://www.webnucleo.org`

We demonstrate that if r-process nucleosynthesis occurs in expansions of matter from high temperature and density in which free nucleons are persistently in disequilibrium with alpha particles, then the resulting abundance distribution can be highly sensitive to the magnitude of strong and electromagnetic reaction cross sections on heavy nuclei. For the particular expansions studied in this work, reactions on nuclei in the atomic number range $Z \approx 40$ and $Z \approx 55$ have the largest effect. These nuclei may be important targets for near-term nuclear cross section experiments.

1. Introduction

The r-process of nucleosynthesis is responsible for production of roughly half the nuclei heavier than iron.[1,2] It is generally thought to proceed through a series of increasingly constrained statistical equilibria (e.g., see Ref. 3), in which case the matter needs to be sufficiently neutron rich to make the heavy neutron-rich isotopes. If, however, expansions of matter from high temperature and density occur rapidly enough, the free nucleons and alpha particles are not in equilibrium with each other during the expansion and it becomes possible to make heavy, neutron-rich nuclei even in matter that is slightly proton rich.[4]

When the nucleosynthesis proceeds through a series of constrained equilibria, the final r-process abundance distribution is quite sensitive to the reactions that assemble heavy nuclei from nucleons and alpha particles but not to reactions on heavy nuclei. In sufficiently fast, high-entropy expansions, however, the possibility of a free-nucleon/alpha-particle disequilibrium can make the abundance distribution highly sensitive to reactions on heavy isotopes. We demonstrate that the magnitude of strong and electromagnetic reaction cross sections on heavy nuclei can govern whether the free nucleons and alphas are in equilibrium. Since these cross sections are generally not known experimentally, they may be important for experimental study in the near future.

2. Alpha-Particle Disequilibrium and Sensitivity to Cross Sections

In order to study the role of reactions on heavy nuclei in fast expansions of high-entropy matter, we ran calculations with the Clemson Nucleosynthesis Code.[5] The network included species from neutrons and protons up to uranium and all relevant strong, electromagnetic, and weak reactions. The matter was taken to be initially slightly neutron rich with a net electron-to-nucleon ratio $Y_e = 0.4975$. The mass density ρ was assumed to evolve with time as

$$\rho(t) = \rho_1 e^{-t/\tau_1} + \rho_2 e^{-t/\tau_2}, \quad (1)$$

where $\rho_1 = 2.265 \times 10^6$ g/cm^3, $\tau_1 = 0.7$ milliseconds, $\rho_2 = 1.195 \times 10^3$ g/cm^3, and $\tau_2 = 100$ milliseconds. The initial temperature was $T_9 = T/10^9$ K $= 10$, which translates to a photon-to-nucleon ratio of 15. This photon-to-nucleon ratio was taken to be a constant during the expansion so that $\rho \propto T^3$. It is useful to note that a photon-to-nucleon of 15 corresponds to an entropy per nucleon of about 150.

The first calculation used standard reaction rates on heavy nuclei.[6] In order to study the effect of the catalyzing effect of the reactions on heavy nuclei, we also ran an identical calculation to the first but this time with all strong and electromagnetic reactions on nuclei with atomic number $Z \geq 20$ increased by a factor of two at all temperatures. In order to ensure that the nuclear populations achieved the correct equilibrium, the code automatically increased the corresponding reverse reaction rates by a factor of two as well.

Figure 1 shows the final abundances in the two calculations as a function of nuclear mass number A. Remarkably, when the reaction cross sections on isotopes of Calcium and heavier are increased by a factor of two, lighter nuclei are produced. The calculation with standard rates produces a robust third r-process peak ($A \approx 195$) while the calculation with the increased cross sections only produces a second r-process peak ($A \approx 130$). The reason for difference this is that described in Ref. 4. Early in these fast, high-entropy expansions, the alpha particles fall out of equilibrium with the free nucleons. This is because the high entropy leads to deuterium, tritium, and ^{3}He abundances that are too low to carry flow to ^{4}He as rapidly as the equilibrium demands. Later in the expansion, however, the alpha particles that are present assemble into heavier nuclei via three-body reactions, such as the triple-α reaction. If enough of these heavy nuclei form, reaction cycles such as ^{56}Ni$(n,\gamma)^{57}$Ni$(n,\gamma)^{58}$Ni$(p,\gamma)^{59}$Cu$(p,\alpha)^{56}$Ni and their reverses can assemble ^{4}He from free neutrons and protons and restore the free-nucleon/alpha-particle equilibrium. If the equilibrium catalyzing reaction cycles are not efficient,

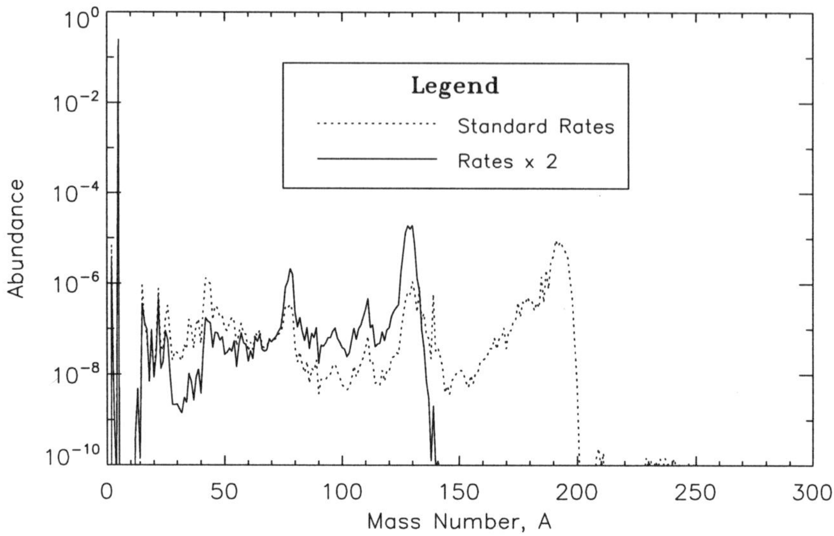

Fig. 1. The final abundances for the calculation with standard rates and the calculation with strong and electromagnetic reaction cross sections for $Z \geq 20$ increased by a factor of 2.

however, the disequilibrium between free nucleons and alphas persists.

For the expansions considered here, the catalyzing nuclear reactions are not fast enough to restore the free-nucleon/alpha-particle equilibrium. However, when we increase the cross sections on Calcium and heavier isotopes, the reactions are faster and do restore the equilibrium. This is evident in Figure 2. The quantity $R_\alpha / R_n^2 R_p^2$ measures how well the free neutrons and protons are in equilibrium with the alpha particles.[7] From Figure 2, it is clear that the free-nucleon/alpha-particle equilibrium fails in the expansion with standard rates near temperature $T_9 \approx 8$ and is never restored. By contrast, in the expansion with the increased cross sections on heavy nuclei, the equilibrium is restored near $T_9 = 5.2$.

It is important to note that, for the high-entropy expansions we consider in this work, the free-nucleon/alpha-particle equilibrium favors ^{4}He over free nucleons as the temperature drops below $T_9 \approx 6$. A disequilibrium between the free nucleons and alpha particles thus leads to a much higher abundance of free neutrons and protons than would prevail if the equilibrium were present. The consequence is that the few heavy nuclei that do form by three-body reactions (namely, the triple-α reaction) capture many neutrons and protons and are thus much heavier than

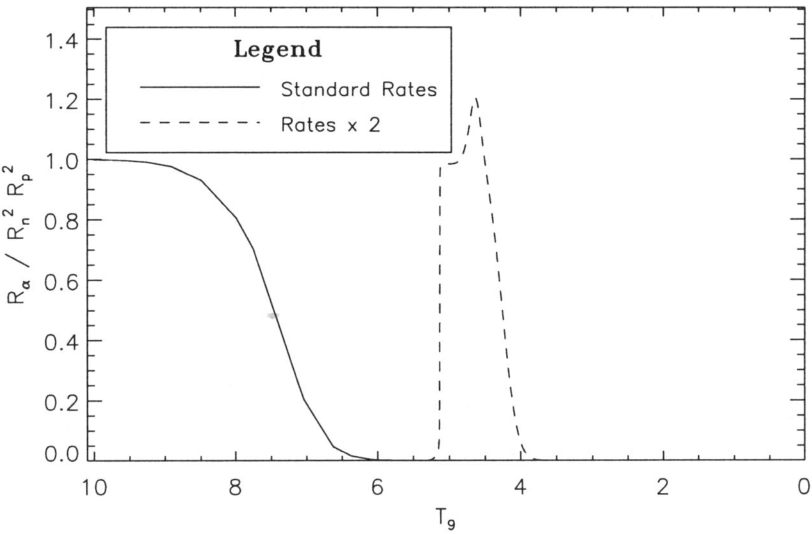

Fig. 2. The quantity measuring equilibrium between free nucleons and alpha particles as a function of temperature for the calculation with standard rates and the calculation with strong and electromagnetic reaction cross sections for $Z \geq 20$ increased by a factor of 2.

they would be if the free-nucleon/alpha-particle equilibrium were present. If the equilibrium is restored, however, the abundance of free nucleons that keeps the nuclei at high mass is depleted and, thus, the heavy nucleus abundance distribution collapses back down towards the iron-group isotopes. Figure 3 shows the number of free protons per heavy nucleus during the two expansions. It is clear that for the expansion with enhanced nuclear cross sections, there are many fewer protons after the free-nucleon/alpha-particle equilibrium is restored. The seed abundance distribution for the r-process phase of the expansion is thus lower in mass than in the expansion with the standard rates. Figure 3 shows that there are also fewer neutrons per nucleus available for the subsequent r-process phase. The result is that the expansion with the enhanced reaction cross sections makes lower-mass nuclei on average than the expansion with the standard rates, as seen in Figure 1.

Since an increase of the strong and electromagnetic reaction cross sections is better able to catalyze the free-nucleon/alpha-particle equilibrium, a decrease of those cross sections should hinder the catalysis and produce heavier nuclei. This is evident in Figure 4. The smaller cross sections leads to a heavier seed distribution

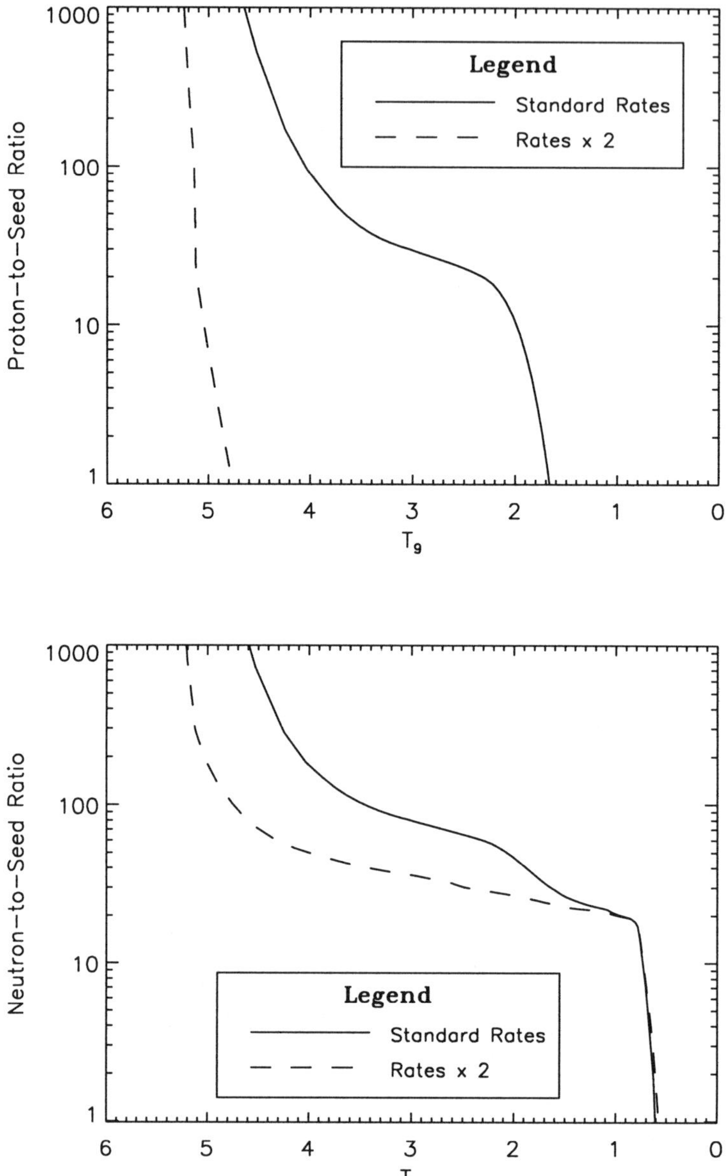

Fig. 3. The number of protons (upper panel) and neutrons (lower panel) per heavy nucleus as a function of temperature for the calculation with standard rates and the calculation with strong and electromagnetic reaction cross sections for $Z \geq 20$ increased by a factor of 2.

and more free particles for the subsequent r-process phase.

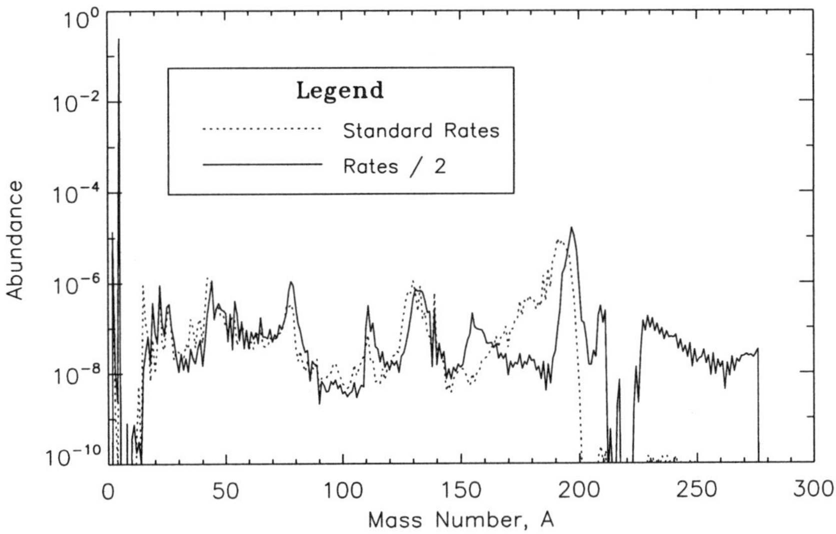

Fig. 4. The final abundances for the calculation with standard rates and the calculation with strong and electromagnetic reaction cross sections for $Z \geq 20$ decreased by a factor of 2.

3. Nuclei of Importance

As Figures 1-4 show, the magnitude of the cross sections for strong and electromagnetic reactions on heavy nuclei can affect the final r-process yields by determining the efficiency of the catalysis of the free-nucleon/alpha-particle equilibrium. In order to get a better sense of which nuclei are important as catalysts for the conditions studied here, we ran identical calculations but with different values of the cutoff atomic number Z_c, that is, the atomic number including and above which we increased the cross sections of strong and electromagnetic reactions by a factor of two. The results of this survey are shown in Figure 5.

Figure 5 shows the average atomic charge $\langle Z \rangle$ and average atomic mass $\langle A \rangle$ of the heavy nuclei in the final abundance distribution as a function of the cutoff Z. There is a sharp jump when Z_c shifts from 39 to 40 and then declines slightly for higher Z_c before rising again. This suggests that nuclei in the atomic number range $Z = 38 - 42$, and particularly with $Z = 39$, are the important catalysts for

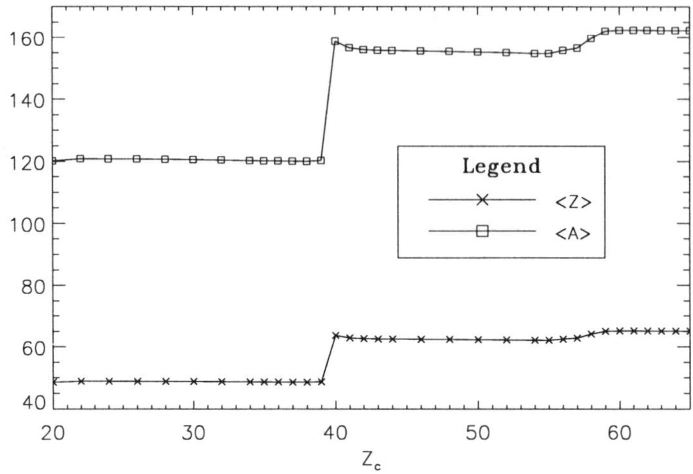

Fig. 5. The resulting average atomic number and mass number of the heavy nuclei when the strong and electromagnetic reaction cross sections for $Z \geq Z_c$ are increased by a factor of 2.

restoring the free-nucleon/alpha-particle equilibrium. It is also evident that cross sections on nuclei in the range $Z \approx 55$ also have an effect on the final abundance distribution.

Figure 6 shows why cross sections on nuclei in the $Z \approx 40$ and $Z \approx 55$ range are important. In the expansion with standard rates, the heavy-element abundance distribution for $T_9 = 5.45$ is dominated by $Z \approx 40$ nuclei. Reactions on these nuclei then are the ones that catalyze the free-nucleon/alpha-particle equilibrium. As the temperature drops, the abundance distribution shifts to even higher charge. At $T_9 = 5.18$ nuclei with $Z \approx 55$ dominate the abundances and, consequently, reactions on these species catalyze the assembly of ^{4}He and deplete the free nucleon abundance.

When the reaction cross sections on nuclei with $Z \geq 20$ are increased by a factor of 2, the reactions on $Z \approx 40$ nuclei restore the free-nucleon/alpha-particle equilibrium. The abundance distribution never reaches the $Z \approx 55$ nuclei. Rather it returns to a quasi-statistical equilibrium dominated by iron-group isotopes. When the reaction cross sections on nuclei with $Z \geq 55$ are increased by a factor of 2, the abundance distribution is able to shift past $Z \approx 40$ up to $Z \approx 55$. The catalyzing reactions on those species do form ^{4}He but not enough to restore the full free-nucleon/alpha-particle equilibrium. The seed dis-

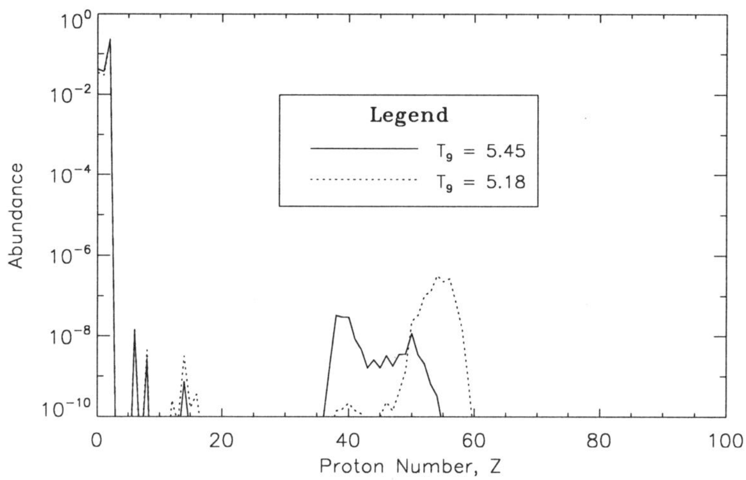

Fig. 6. The elemental abundances at $T_9 = 5.45$ and $T_9 = 5.18$ in the calculation with standard rates.

tribution is a little lower in mass and the abundance of free neutrons is a little less than in the calculation with standard rates. This is the cause of the slight rise in the average nuclear mass and atomic number in Figure 5 when Z_c is above $Z \approx 57$. The reader is invited to study this further by viewing the movies at `http://nucleo.ces.clemson.edu/home/movies`.

The specific reaction pathways that are important for catalyzing the free-nucleon/alpha-particle equilibrium are difficult to determine because they involve many nuclei simultaneously. Nevertheless, it is clear that the important isotopes are beta-stable nuclei or near stability. Figure 7 shows the nucleosynthesis "path", that is, the locus of most abundant isotopes for each element, at $T_9 = 5.18$. It is evident that at this temperature, the most abundant isotopes are near stability. Only at lower temperatures does the path move to more neutron-rich species. This again is best seen in movies at the URL above.

4. Discussion

Recent work suggests that supernovae can indeed achieve conditions in which nucleosynthesis occurs with a persistent free-nucleon/alpha-particle disequilibrium.[8] If indeed this result is further confirmed, experimental studies of strong and electromagnetic reaction cross sections on nuclei in the $Z \approx 40$ and $Z \approx 55$ regions

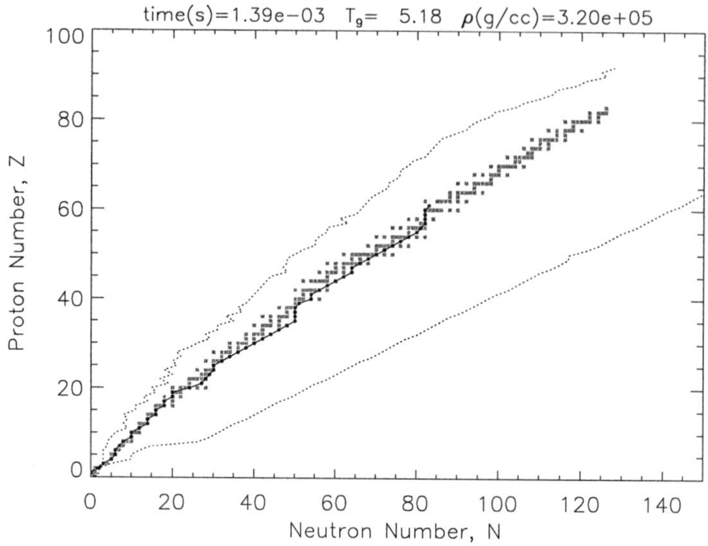

Fig. 7. The r-process "path", that is, the locus of most abundant isotopes for each element, for $T_9 = 5.18$ in the calculation with standard rates. The stable isotopes are the small solid boxes. The limits of the network are also shown.

will become important for good estimates of the resulting nuclear yields. Interestingly, the important isotopes will be near beta stability and, hence, can be studied with facilities currently existing or planned for construction in the near future.

Acknowledgments

The authors gratefully acknowledge support from NASA's Cosmochemistry Program and DOE SciDAC grant DE-FC02-01ER41189.

References

1. E. M. Burbidge, G. R. Burbidge, W. A. Fowler and F. Hoyle, *Rev. Mod. Phys.* **29**, 547 (1957).
2. A. G. W. Cameron, *Pub. Astron. Soc. Pacific* **69**, 201(June 1957).
3. B. S. Meyer, G. C. McLaughlin and G. M. Fuller, *Phys. Rev. C* **58**, 3696(December 1998).
4. B. S. Meyer, *Physical Review Letters* **89**, p. 231101(November 2002).
5. G. C. Jordan and B. S. Meyer, *Astrophys. J.* **617**, L131(December 2004).
6. T. Rauscher and F. Thielemann, *At. Dat. Nucl. Dat. Tab.* **75**, 1(May 2000).
7. B. S. Meyer, T. D. Krishnan and D. D. Clayton, *Astrophys. J.* **498**, 808(May 1998).
8. F. X. Timmes, Fryer, C. L., A. Hungerford and F. Herwig, *Astrophys. J.* (2006), submitted.

PREDICTIVE R-PROCESS CALCULATIONS

C. L. FRYER,* F. HERWIG, A. L. HUNGERFORD, F. X. TIMMES
Los Alamos National Laboratory
Los Alamos, NM, 87544, USA
E-mail:fryer@lanl.gov, fherwig@lanl.gov, aimee@lanl.gov,timmes@lanl.gov

Understanding the nature of the r-process is one of the great unsolved problems in astrophysics. Obtaining accurate nuclear rates will eventually be critical in solving this problem. But to truly take advantage of accurate rates, we must understand the r-process beyond the phenomenological level of postdictive science and move toward predictive models based on calculations anchored by physical models. Using the fallback r-process site, we outline the process of identifying the aspects of this model that are not yet understood and assessing the uncertainties in our lack of understanding. These are the first 2 steps of predictive modeling that can lead to the final step: addressing the aspects that produce the largest uncertainty and minimizing that uncertainty.

1. Predictive Science and the r-Process

The understanding of any astrophysical phenomena progresses through a series of stages. The discovery of any new phenomenon is generally immediately followed by a rough, back-of-the-envelope calculation which really answers the questions: is it in the realm of possibility that we can explain this new phenomena. In supernova observation theory, Arnett[1,2] came up with such a back-of-the-envelope calculation for the flux produced by a supernova event as a function of time. This theory showed that the explosion of a massive star might explain the observational phenomenon known as a supernova. Arnett's model also could explain the basic features of supernova light curves and his models suggested that the difference between type I and type II supernovae are the size of the stars (something most supernova researchers now take as fact).

The second stage is an enhanced version of the first. It develops a model that can give quantitative answers. Generally many free parameters are introduced to allow the would-be theorist to match the observations well. In supernova theory, an example of such a "theory" is the SYNOW code developed by Fisher et

*c.l. fryer and f.x. timmes are also at the physics department, univ. arizona

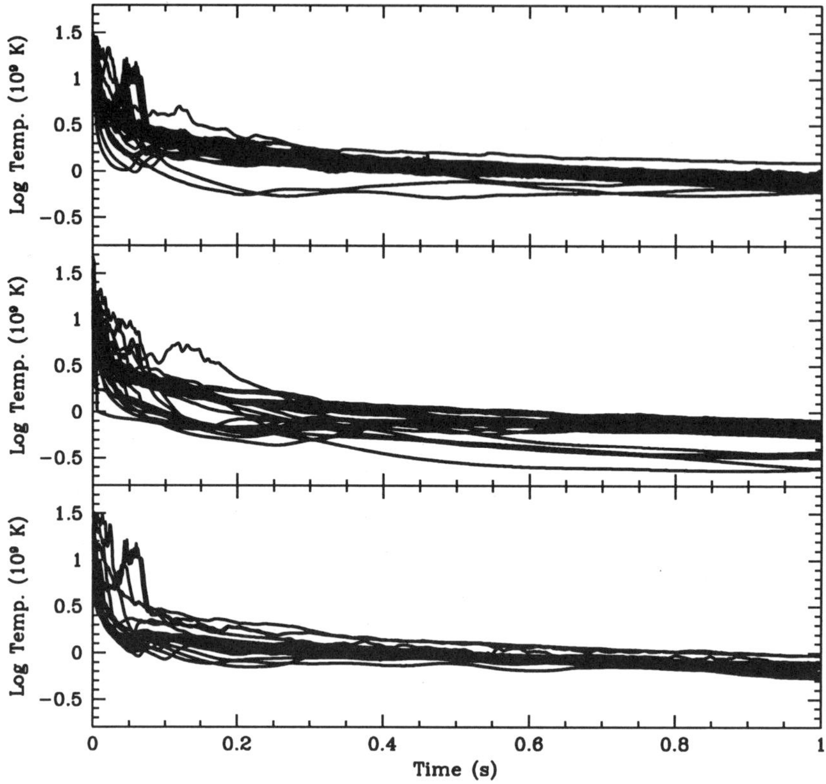

Figure 1. Temperature versus time for 3 sets of particles: high entropy (top), low entropy (middle) and particles that produced reasonable amounts of the third r-process peak (bottom). The zero point on the time axis is set to the time when the density reaches its maximum value. This generally corresponds to the peak in temperature as well.

al[3]. SYNOW stands for synthetic spectrum now, meaning that the code is simple enough that one can run it quickly (1-2 hours) on a single-processor machine. Among its free parameters is the abundance distribution of elements and it is used to provide a rough guess for the elements present in an exploding star. Although this model is based on basic principals of physics, it is not a truly physical representation of nature. Indeed, its free-parameters will allow a theorist to fit an observation with an unphysical set of conditions. But such a code serves a very important use in classifying supernovae. In part, its usefullness arises from its practitioners understanding its limitations.

The final stage is to build a model entirely based on physics. Such a model will also have uncertainties (e.g. in the initial conditions, in the implementation of the physics), but these uncertainties are based on real physical issues based on trying to do detailed calculations using a computer. In our supernova light-curve example, the uncertainties arise either because models tend to either assume local thermodynamic equilibrium (or something near to it) or time independence. Correction factors sometimes arise to try to address the limitations of these assumptions, but generally these assumptions are considered to be the subject of further study on the road to the "correct" answer. In these models, the exact uncertainties in the opacities are important. These calculations, without free parameters, will always fit the data worse than codes like SYNOW. But these calculations will ultimately allow supernova theorists to predict (not just postdict) what a supernova will look like. In the supernova game, the entire community, including observers, realizes that SYNOW does not provide the final answer and the work of predictive science using first-principal calculations is crucial.

With the r-process problem, we are still in the second stage of supernova theory. A simple model, the "wind" model has been developed and, if one allows its free parameters to be tuned to unphysical levels, it can match the r-process yields exactly[4,5,6]. Unfortunately, this accurate fit has allowed r-process theory to degenerate into a "dark-age" where very little progress has occured since the seminal papers by Qian, Woosley and collaborators. Scientists studying the r-process have been wary of moving to the third stage. This is because working in the third stage will almost certainly produce fits that are not as good as what we can do with parameter models. But to get beyond the phenomenological phase, we must move to physical models. We argue here that getting better nuclear reaction rates is also pointless until we move on to this third stage.

The wind model works, but it requires unphysically high entropies to do so. What this should have told the r-process community is that this model really *doesn't* work. Fortunately, other formation sites for the r-process exist. One such model is the fallback scenario[7]. In this scenario, material falling back onto the proto-neutron star is heated to produce heavy elements. A fraction of this fallback is ejected, and Fryer et al.[7] have shown that it can make the r-process elements. In addition, these authors did not develop a parameter model so if the uncertainties in this model can be minimized, this model *can* take advantage of the progress in nuclear reaction rates. In this paper, we outline our current progress in analyzing the uncertainties in this model: the two primary uncertainties being the particle trajectories and the evolution of the electron fraction.

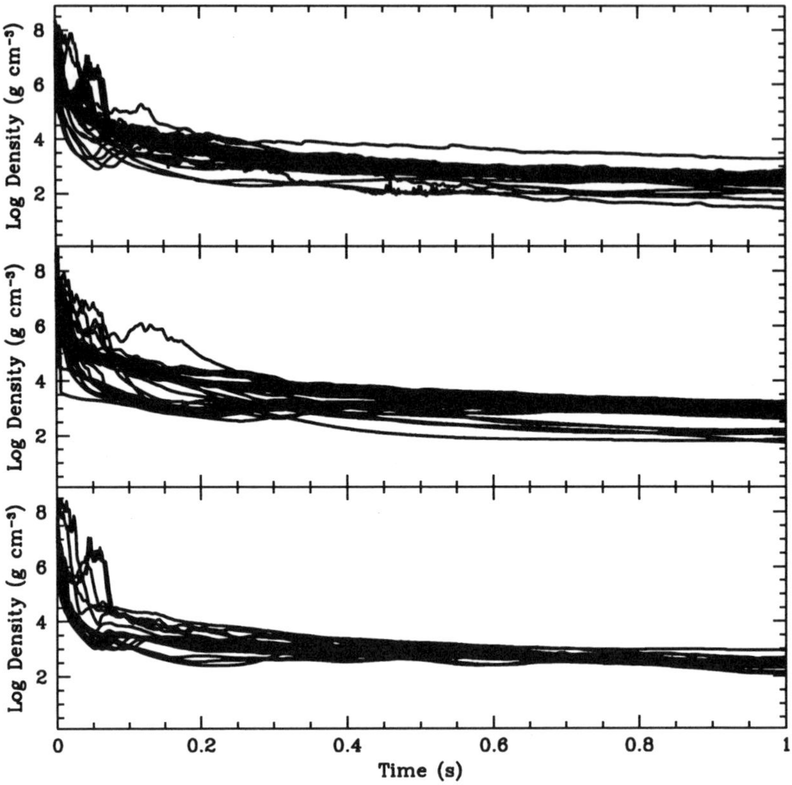

Figure 2. Density versus time for the same 3 sets of particles as plotted in Figure 1. As with the temperature profile, there are no obvious features displayed in those particles that produce the third r-process peak.

2. Fallback Trajectories

It has long been known that most supernovae produce some fallback (see Ref. 8 for a review). Predictive simulations of core-collapse and fallback (this third stage of theory) allowed Fryer & Kalogera[9] to argue that the observations circa 2001 that predicted a single mass for nascent neutron stars and a second single mass for all nascent stellar black holes must be wrong. It is now known that the neutron stars and black holes are formed with a range of masses and the observational community has acknowledged their errors in interpreting the data dur-

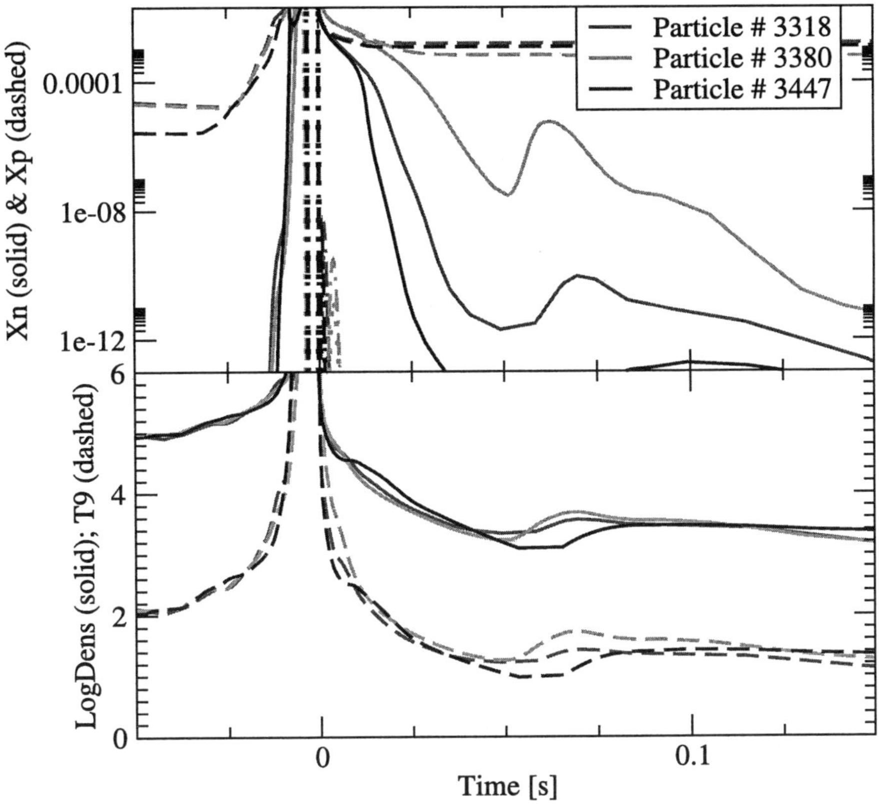

Figure 3. The density (solid) and temperature (dashed) profiles as a function of time for 3 specific particles that produced a range of heavy r-process elements (bottom panel). The top panel shows the neutron (solid) and proton (dashed) fractions for these same particles as a function of time. Clearly, the free neutron fraction as a function of time is very sensitive to the deviations in the density and temperature profiles. And it is the capture of these neutrons that produce the r-process yields we observe.

ing that time. This example shows the strength of a predictive model. A truly predictive model can cut through any bias in the interpretations of observations, ultimately allowing us to better understand what we are seeing in the data. This predictive success shows that fallback is indeed an important feature required to match the observational data.

This fallback material rains down onto the proto-neutron star releasing an enormous amount of gravitational potential energy. This energy can be used to eject some of the fallback mass. It is this ejected mass that Fryer et al.[7] studied as a possible source of the r-process. Fryer et al.[7] identified the features required for this ejecta to produce heavy elements. The temperature profile initially drops very quickly with time (allowing the protons and neutrons to fall slightly out of equilibrium with the alpha particles). Then the temperature decay must slow down significantly to allow protons to build up heavy elements (bypassing waiting points) with continued neutron capture to build neutron rich isotopes. Such a path had already been discovered by Brad Meyer[10]. But the exact conditions for making heavy elements requires more than a two-sloped density/temperature profile. We know this because nearly all of the trajectories in the Fryer et al.[7] model have roughly this profile, but only about 3% of all particles make the third r-process peak.

To try and distinguish the feature that is key for making r-process elements, we plot the temperature evolution of the top 3% r-process producing particles (Fig. 1 - bottom panel). The top panel in Fig. 1 shows the temperature evolution of the high entropy particles and the middle panel shows the same evolution of the low entropy particles. The third-peak producing particles have trajectories that are more similar to the high-entropy particles (indeed, 25% of all the third-peak producing particles are high entropy), but entropy is not the sole determiner of heavy element r-process production. Not all high entropy particles produce massive r-process elements, and some of the particles that produce the third-peak in the r-process are relatively low entropy.

Fig. 2 shows the same data for the density evolution. We see the same trends (all exhibit a fast drop followed by a slow decay, the third-peak producing particles more closely match the high-entropy particles, etc.) as those seen in our temperature evolution, but there is no clear factor that is key for heavy element production. Indeed, if we look at the trajectories of three particles that produced varying amounts of the third r-process peak (Fig. 3), it is difficult to see how the temperature/density evolution differences and similarities can dramatically change the yield.

3. Electron Fraction

Fryer et al.[7] used a very simple picture for the time evolution of the electron fraction (Y_e). First, in their hydrodynamic models, they assumed $Y_e = 0.45$. This electron fraction then evolved with time, slowly increasing (Fig. 4). But in their post-process step to calculate the nucleosynthetic yield, they assumed $Y_e =$

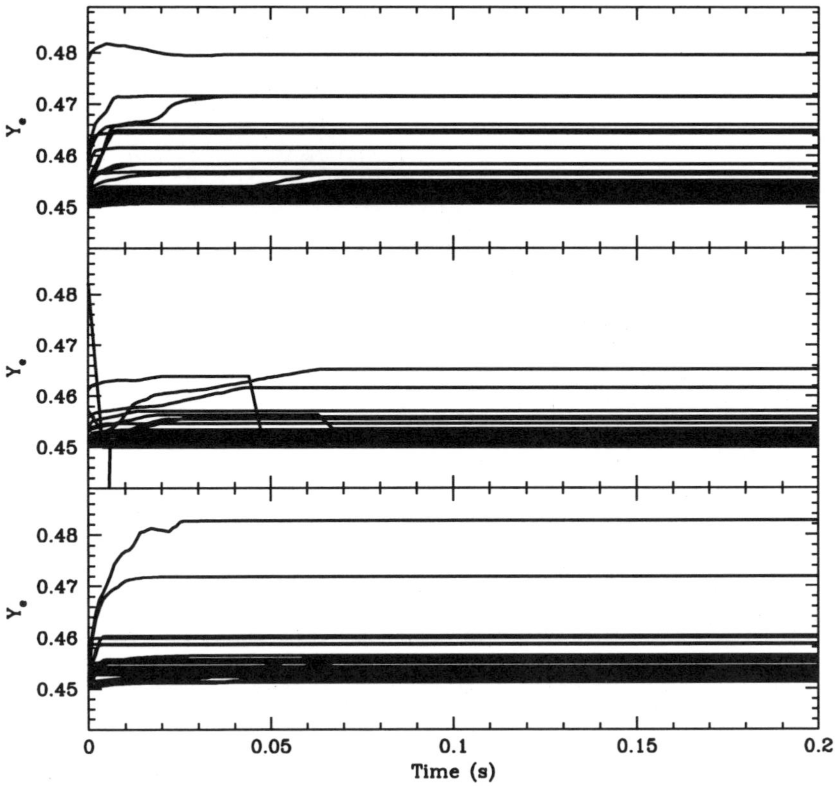

Figure 4. Electron fraction (Y_e) versus time for the same 3 sets of particles as plotted in Figure 1.

0.5. The hydrodynamic evolution is fairly insensitive to the electron fraction, so a hydrodynamic calculation assuming $Y_e = 0.5$ would not produce significantly different results than one of $Y_e = 0.45$.

But the nucleosynthetic yield will vary significantly with the exact value of the electron fraction. Figure 5 shows a yield plot (shaded by abundance fraction) for our $Y_e = 0.5$ simulation. The abundance of material lies just below the valley of stability. This will produce r-process elements, but not with the exact positions observed in nature. The parameter-theory trained observer would then claim that such a yield must be incorrect. This is because theory that has many parameters

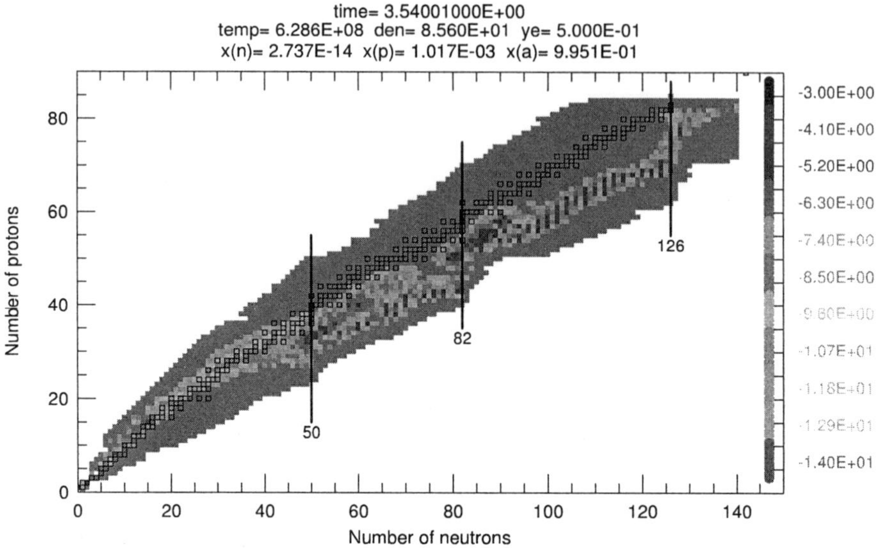

Figure 5. A yield plot showing the fraction of elements (color-coded) on a 2-d plot showing the neutron number (x-axis) and proton number (y-axis). The yield is clearly neutron-rich, but it has a slightly different structure than the classic r-process peaks. As we see in figure 6, this yield is very sensitve to the electron fraction.

(the second stage of theoretical astronomy) will match the data exactly. In the physics-based theory we describe here, we do not expect the theory to match the data exactly. In fact, we argue that any theory that does match the data exactly must certainly be wrong, as there are uncertainties in the theory (e.g. nuclear rates) that will change the final answer (see Ref. 11 for example).

Alternatively, a change in the electron fraction can also change the yield. Fryer et al.[7] ran simulations with constant electron fractions slightly modified away from a value of $Y_e = 0.5$. Just 1% variations in the electron fraction can completely change the yield (Fig. 6).

4. Hope for the Future

We have presented a new model for r-process, one based on physics with no parameters. This truly is a predictive theoretical model. Like any physical model, we do not tune this model to produce the exact yields of any observation. But there are uncertainties, in the electron fraction, in the initial conditions (angular momentum, fallback rate) that will change the particle trajectories, and in the nu-

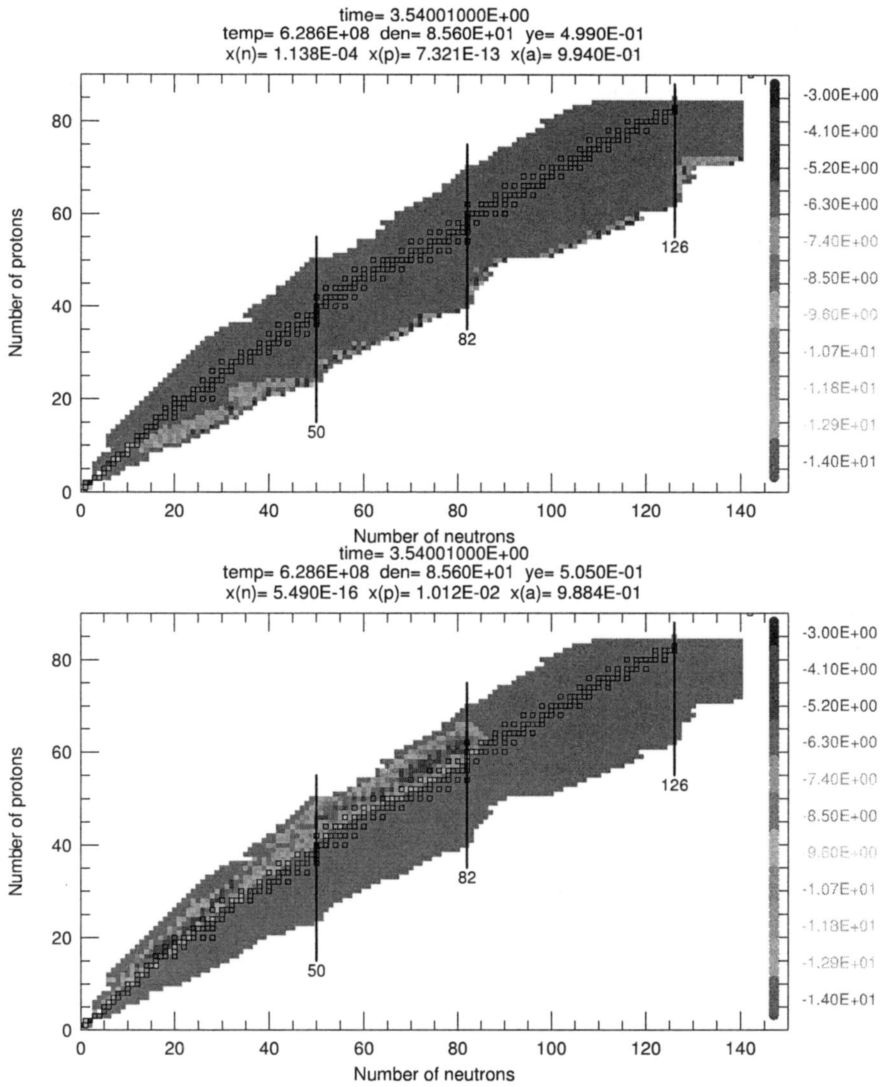

Figure 6. Two different yield plots for models with ±1& variance around the $Y_e = 0.5$ yield plot in Figure 5. The low electron fraction $Y_e = 0.495$ model produces a neutron rich yield that is similar to classic r-proces. The $Y_e = 0.505$ model produces a very different proton-rich yield.

clear reaction rates. We do not doubt that we can fine-tune these to produce whatever yields we want. But we argue that the more important work on this problem should focus on understanding these uncertainties, not sweeping them under the rug with a few well-placed parameters. It is by understanding them that we will ultimately be able to believe with confidence the true affect the nuclear rates have on the r-process yield.

References

1. Arnett, D. 1979, ApJ, 230, L37.
2. Arnett, D. 1980, ApJ, 237, 541.
3. Fisher, A., Branch, D., Nugent, P., & Baron, E. 1997, ApJ, 481, L89
4. Duncan, R.C., Shapiro, S.L., & Wasserman, I. 1986, ApJ, 309, 141
5. Qian, Y.-Z., & Woosley, S.E. 1996, ApJ, 471, 331
6. Hoffman, R.D., Woosley, S.E., & Qian, Y.-Z., 1997, ApJ, 482, 951
7. Fryer, C.L., Herwig, F., Hungerford, A.L., & Timmes, F.X., submitted to ApJ, (2006).
8. Fryer, C.L. 1999, ApJ, 522, 413
9. Fryer, C.L. & Kalogera, V. 2001, ApJ, 554, 548
10. Meyer, B.S., PRL, 89, 1101 (2002).
11. Engel, J., Bender, M., Dobaczewski, J., Nazarewicz, W., and Surman, R. 1999, PRC, 60, 014302

NUCLEI AS LABORATORIES: NUCLEAR TESTS OF FUNDAMENTAL SYMMETRIES

M. J. RAMSEY-MUSOLF

California Institute of Technology, Pasadena, CA 91125, USA

The prospect of a rare isosotope accelerator facility opens up possibilities for a new generation of nuclear tests of fundamental symmetries. In this talk, I survey the current landscape of such tests and discuss future opportunities that a new facility might present.

1. Introduction

The mission of nuclear physics in the coming decade is to explain the origin, evolution, and structure of the baryonic matter in the Universe[1]. Historically, nuclear physics has played a key role in developing our Standard Model (SM) of particle physics through exquisite tests of fundamental symmetries such as parity (P) and time-reversal (T) invariance. Today, we are on the cusp of a new era in particle physics, with the imminent operation of the Large Hadron Collider (LHC). One hopes that the LHC will both uncover the mechanism of electroweak symmetry-breaking associated with the yet unseen Higgs boson and discover evidence for physics beyond the SM. At the same time, cosmology points to the need for new particle physics, as the SM fails to account for the abundance of visible, baryonic matter in the Universe, the existence of cold dark matter, and the mysterious dark energy responsible for cosmic acceleration. The corresponding challenge for particle and nuclear physics is to determine what physics beyond the SM can explain these cosmic building blocks and their relative contributions to the energy density of the Universe.

Within this context, the focus for nuclear physics falls squarely on the baryonic component. It is up to nuclear physics to explain how baryonic matter came to be in the first place; how it evolved as the Universe cooled; how it coalesced into the protons and neutrons and other hadrons; how the fundamental quarks and gluons are put together inside baryons; how their interactions give rise to atomic nuclei in all their rich complexity; and how these nuclei and their interactions drive the formation, structure, and life cycle of stars. In pursuing this mission, it is essential to understand how the basic forces of nature have shaped this story of the baryons. Despite the tremendous successes of nuclear physics over the decades, our knowledge of this story remains quite limited. We are able to account for many – but not all – properties of baryonic matter only after quarks and gluons coalesced into protons and neutrons. We don't know, however, whether the strong, electroweak, and gravitational forces that dominate at late times were the only or even the most important ones at times before the formation of QCD bound states. We don't know whether the interactions of quarks with leptons looked the same at earlier times as they appear today. And we certainly have no definitive explanation of how the tiny – but anthropically relevant – excess of baryonic over anti-baryonic matter arose after the initial blast of the Big Bang.

In this talk, then, I want to discuss how we can exploit atomic nuclei as "laboratories" to look back in time to try and arrive at insights into these questions. In doing so, we have to exploit what we know about the symmetries of SM interactions and – by performing precise tests of these symmetries – piece together clues about the interactions of baryonic matter at early cosmic times. At the same time, there remain a host of poorly understood features of baryonic matter in the present era, and among the most enigmatic, is the weak interaction between quarks. So I will divide this talk into two broad questions:

i) *What were the fundamental forces that governed the interactions of quarks in the early Universe?* In particular, we would like to determine what forces were responsible for the generation of the excess of baryonic matter as well as what other forces shaped the dynamics of quarks once they were created. In addressing this question, I will discuss how searches for the permanent electric dipole moments (EDMs) of the neutron and atoms and precision studies of weak nuclear decays can provide important insights. In each case, the violation of fundamental symmetries – including P, T, and C – provide essential handles in probing the forces of the early Universe.
ii) *How do the electroweak and strong interactions of the SM shape the weak interactions of baryons in the present Universe?* Because the strong and electromagnetic interactions are so much more powerful than the weak quark-quark interaction at the low-energies relevant to nuclear physics, we must exploit the P-violating (PV) character of the weak interaction to observe it experimentally. I will discuss how a new generation of hadronic PV experiments are poised to provide a new window on the hadronic weak interaction (HWI) and how – ultimately – these studies may help is better understand the nuclei as laboratories as per point (i) above.

Before proceeding, I wish to acknowledge that I will be forced to omit some important topics in nuclear physics tests of fundamental symmetries and will undoubtedly and unintentionally miss some important references to recent work. I hope that these omissions will be remedied by other recent reviews that I have given[2,3] , and I refer the interested reader to those discussions.

2. Electric Dipole Moments and the Origin of Baryonic Matter

It is widely believed that the initial, post-inflationary conditions were matter-antimatter symmetric. If so, then the particle physics of the post-inflationary era would have to be responsible for generating a nonvanishing baryon number density, n_B. Expressed as a ratio to the photon entropy density at freeze out, s_γ, the baryon asymmetry is

$$Y_B \equiv \frac{n_B}{s} = \begin{cases} (7.3 \pm 2.5) \times 10^{-11}, & \text{BBN}^{\,5} \\ (9.2 \pm 1.1) \times 10^{-11}, & \text{WMAP}^{\,6} \end{cases} \tag{1}$$

where the first value (BBN) is obtained from observed light element abundances and standard Big Bang Nucleosynthesis and the second value is obtained from the cosmic microwave background as probed by the WMAP collaboration. Forty years ago, Sakharov identified the three key ingredients for any successful accounting for this number[4]: (1) a violation of baryon number (B) conservation; (2) a violation of both C and CP symmetries; and (3) a departure from thermal equilibrium at some point during cosmic evolution[a].

In principle, these ingredients could have generated $Y_B \neq 0$ at any moment in the post-inflationary epoch up to the era of electroweak symmetry breaking. At one extreme, baryogenesis might have occurred at very early times, associated with particle physics at scales much greater than the electroweak scale, $M_{\rm wk}$. At the other end is the possibility of electroweak baryogenesis (EWB). And, Y_B may have been generated at some point between these two cosmic "bookends". During the coming decade, experiments that probe new weak scale physics, including EDM searches and the LHC studies, will test EWB with revolutionary power. In the most optimistic scenario, these experiments will uncover the building blocks of EWB and point to the new physics scenario that consistently incorporates them. Even null results, however, would be interesting, as they would imply that EWB is highly unlikely and point to higher scale scenarios – such as GUT baryogenesis or leptogenesis – that are more difficult to test experimentally. In the case of leptogenesis one can at least look for some of the elements that have to exist to

[a]The last ingredient can be replaced by CPT violation, and some baryogenesis scenarios exploit this fact.

make this scenario plausible: CP-violation in the lepton sector, an appropriate scale for m_ν; and lepton number violation. Neutrino oscillation studies – together with ordinary β-decay and neutrinoless double β-decay ($0\nu\beta\beta$) – provide our means for doing so[b]. Below, I will comment on $0\nu\beta\beta$ in more detail.

In the case of EWB, it is natural to ask why the SM is insufficient since the SM contains all of Sakharov's criteria:

(1) *Baryon number violation.* In the SM takes place in the through anomalous processes called sphaleron transitions. The difference of baryon lepton number currents, $j^B_\mu - j^L_\mu$, is anomaly-free and conserved in the SM, but their sum is not. There exist an infinite number of vacua differing in total $B + L$ by integer units, and the probability for tunneling between these $T = 0$ vacua is highly suppressed. At temperatures of order the electroweak scale ($T_{\rm wk}$), however, the thermal excitations of gauge configurations with energy of order $T_{\rm wk}$ can occur with finite probability. These configurations can decay to a different vacuum than the one out or which they were created, thereby changing $B + L$.

(2) *Departure from thermal equilibrium.* In principle, the generation of mass in the SM at the electroweak scale could have provided the necessary departure from thermal equilibrium through a transition between the phases of unbroken and broken electroweak symmetry. In order to ensure that any non-zero n_B created by sphaleron processes is frozen into the broken phase (where we live), this phase transition needs to be strongly first order. The parameters of the Higgs potential are critical in determining whether or not such a first order phase transition can occur. Given the present lower bounds on the Higgs boson mass obtained from LEP II ($m_H > 114.4$ GeV) the parameters of the Higgs potential that depend on this mass and on the weak scale cannot lead to a strong first order phase transition. New physics that couples to the Higgs sector is needed to bring about such a phase transition.

(3) *CP-violation.* The presence of CP-violation is needed to generate a net asymmetry between production of left- and right-handed particle densities, and it is this imbalance that feeds the $B + L$ violating sphaleron processes[c]. The electroweak sector of the SM contains CP-violation via the phase in the CKM matrix that affects interactions between the

[b]It should be emphasized, however, that the absence of experimental evidence for lepton C- and CP-violation does not preclude leptogenesis.

[c]This chiral charge production can be indirect, taking place first via generation of a net Higgs number density and then transferring to the fermion sector through Yukawa-like interactions.

spacetime varying Higgs vev and quarks at the phase transition boundary. Unfortunately for the SM, the net effect of this CP violation is highly suppressed by the Jarlskog invariant, J, and multiple powers of quark Yukawa couplings[7d]. Consequently, viable EWB requires the presence of new CP-violating interactions beyond those of the SM that do not suffer from this suppression. In the absence of such Jarlskog suppression, one would also expect the effects of these CP-violating interactions to enter more strongly in EDMs than does SM CP-violation.

As reviewed in Ref. [9], there exist a number for models for new physics at the weak scale that can remedy these SM shortcomings. A strong, first order phase transition naturally arises in supersymmetric models, for example, as they contain new scalar degrees of freedom – such as the scalar superpartners of fermions or singlet Higgs fields – that couple appropriately to the SM-like Higgs and correspondingly modify the potential. These supersymmetric interactions can also include new CP-violating effects that are not suppressed as in the SM, thereby allowing for the requisite creation of chiral charge. Whether or not such models can lead to successful EWB depends on the characteristics of the Higgs potential, the details of quantum transport at the phase boundary, and on the values of the model parameters.

In addressing these issues, improvements in both theory and experiment are important. Theoretically, refined treatments of quantum transport, the details of the Higgs potential, and the dynamics of the expanding regions of broken electroweak symmetry ("bubbles") are being pursued by our group and several others[10]. Experimentally, searches for new physics at the LHC, precision electroweak tests, and both EDM and dark matter searches are poised to provide critical new information about the shape of any new physics at the electroweak scale. From the standpoint of EWB, the EDM experiments will be essential in nailing down the parameters associated with new CP-violation. Moreover, experiments carried out with different systems – such as the electron, muon, neutron, neutral atoms, and even the deuteron – can provide complementary information. If, for example, a non-zero EDM is observed in one such system, there will exist a variety of possible models that can explain it. The results of experiments in other systems will be needed to sort out among the competing explanations.

I would like to emphasize that the possibilities for new, neutral atom EDM searches using rare isotopes are quite compelling in this respect, and they could

[d] Farrar and Shaposhnikov subsequently argued that the SM baryon asymmetry is suppressed only by J and $m_s - m_d$[8].

remain powerful probes of new CP-violation relevant to baryogenesis well into the LHC era. To illustrate, consider EWB in the Minimal Supersymmetric Standard Model (MSSM). The CP-violating phases relevant to MSSM EWB are already highly constrained by present limits obtained from EDM searches if the masses of all the supersymmetric particles participating in one-loop contributions to the EDMs are below about one TeV. It is possible to relax the EDM constraints if the masses of the sleptons and squarks (the scalar, supersymmetric partners of the leptons and quarks) are allowed to become heavy – on the order of 3-10 TeV (see, *e.g.*, Ref. [11] and references therein). In this case, the EDMs of the electron and neutron are dominated by two-loop graphs that involve virtual, SM particles and the superpartners of the electroweak gauge and Higgs bosons – the gauginos and Higgsinos, respectively. The situation with diamagnetic neutral atoms is somewhat different, as the neutral atom EDMs are dominated by long range, CP-violating forces the nucleus that arise from the so-called "chromoelectric" dipole moment (CDM)[12]. In the limit of heavy sfermions, the one-loop CDMs are suppressed just like the one-loop EDMs. However, there are no two-loop CDM contributions associated with virtual gauginos and Higgsinos unlike the EDM case. Thus, if nature had selected this variant of SUSY, one could expect to see non-vanishing electron and neutron EDMs that are consistent with the CP-violation needed for EWB but vanishing atomic EDMs in systems such as Xe or Ra.

3. Weak Decays

The study of weak decays of hadrons has an illustrious history in nuclear physics and has provided key input for the development of the SM. The classic experiment on ^{60}Co by Wu *et al.* [13] that provided evidence for parity violation in the weak interaction – followed by the analogous demonstration in polarized μ^+ decay[14] – pointed the way to our understanding of the $(V-A)\otimes(V-A)$ structure of the SM CC interaction at low energies. Similarly, the comparison of Fermi constants extracted from muon and β-decay lead to the development of Cabibbo mixing between $\Delta S = 1$ and $\Delta S = 0$ charged currents and to notation of that weak and strong interaction eigenstates of quarks are not identical – precursors of the full CKM model for quark mixing. Today, superallowed nuclear β-decays provide the most precise determination of any element of the CKM matrix, namely, V_{ud} – a triumph of precision nuclear physics that has been recognized by last year's award of the Bonner Prize to Hardy and Towner.

That history of accomplishment notwithstanding, one might ask what relevance nuclear β-decay studies will have in the coming LHC era. From my perspective, the answer is all about precision. To the extent that one can push the experimental sensitivity of various β-decay studies significantly beyond levels, this use of nuclei as laboratories will provide powerful probes of new forces at the weak scale that can complement what we may learn from the LHC or even a subsequent e^+e^- collider. The way β-decay studies do so is to search for tiny deviations from SM predictions for various quantities, such as half lives or decay correlation parameters. The pattern of such deviations – or of their absence – can provide new clues about the character of new physics at the electroweak scale if it exists and in some cases test elements of new physics models that are more difficult to access with colliders. The key, however, is to carry out these experiments with greater degrees of precision than before – a prospect that a new rare isotope facility, together with the new fundamental neutron physics beam line at the Spallation Neutron Source – make possible. Let me illustrate this idea with two kinds of β-decay tests.

β-decay and CKM unitarity. The CKM matrix in the SM is unitarity, so any apparent departure from this unitarity property can point to new physics. The most precise such test involves elements of the first row, including V_{ud} obtained from superallowed nuclear β-decays; V_{us} that is obtained most precisely from kaon leptonic (K_{e3}) decays; and V_{ub} determined from B-meson decays. The value of V_{ub} is too small to be relevant to such tests, given the level of precision with which the other two first row elements are known, so I will focus on them. As is well-known to people at this meeting, there had been a $\gtrsim 2\sigma$ deviation from first row unitarity for many years, and it was long thought by many outside our field that the culprit lay in some poorly understood feature of nuclear structure. I have always found that objection hard to swallow in light of the impressive agreement between corrected ft values among the various decays. If the CVC property of the SM is valid, then the corrected ft values for all superallowed decays should be identical. The analysis of the twelve best measured cases indicates that such agreement occurs at the level of a few parts in 10^4[15]. Since the unitarity deficit had been at the part per thousand level, it is hard to see how nuclear structure effects could be the reason in light of the almost order of magnitude more precise agreement with CVC among the twelve superallowed cases of interest.

Recently, new measurements of K_{e3} decays have lead to a shift in the world average for the branching ratios (for a recent summary, see Ref. [16]). The extraction of V_{us} from these branching ratios, however, depends on knowledge of a K-to-π form factor, $f_+(q^2 = 0)$, and there currently exists some controversy over the value of this quantity. It can be analyzed in chiral perturbation theory (ChPT) and all contributions through order p^4 are known. These contributions include both one-loop corrections and contributions from tree-level $\mathcal{O}(p^4)$ operators whose coefficients are known from other experiments. To perform a test of the CKM matrix at the 0.1% level, however, the $\mathcal{O}(p^6)$ contributions are also required. The loop contributions (that include both one- and two-loop corrections) have been computed[17,18], but the tree-level $\mathcal{O}(p^6)$ operator contributions are not known in a model-independent way. Values for these quantities have been obtained using large-N_C QCD methods[19] and lattice QCD computations[20,21,22,23]. The results of the two approaches point in contradictory directions: the large-N_C value for $f_+(0)$ would yield a value for V_{us} that – in combination with the nuclear result for V_{ud} – disagrees with unitarity at the historically irritating $\sim 2\sigma$ level; the quenched lattice results, in contrast, point to unitarity agreement. It will be quite interesting to see how this situation settles down in the next few years as hadron structure calculations improve. In the meantime, new tests of the nuclear structure corrections that are applied to the experimental ft values are being carried out in regions of the periodic table where they are expected to be larger. Future studies at a rare isotope facility would presumably be of interest in this respect.

The implications of unitarity tests for new physics at the weak scale can best be understood by considering the relationship between the vector Fermi constant, G_V^β, that governs nuclear decays with the Fermi constant, G_μ, obtained from the muon lifetime. One has

$$G_V^\beta = G_\mu \, V_{ud} \, (1 + \Delta\hat{r}_\beta - \Delta\hat{r}_\mu) \quad , \tag{2}$$

where $\Delta\hat{r}_{\beta,\mu}$ are $\mathcal{O}(\alpha/4\pi)$ electroweak radiative corrections to the two decay amplitudes. The quantity $\Delta\hat{r}_\beta$ contain hadronic structure uncertainties associated with the $W\gamma$ box graphs involving the lepton and nucleon pairs. Recently, Marciano and Sirlin have reduced this uncertainty by relating the short distance part of the one-loop integral to the Bjorken sum rule and by applying large N_C correlators to the pertrubative-nonperturbative transition region[24]. The resulting value for V_{ud} obtained by these authors and by an up-dated global analysis of superallowed decays[25] is $V_{ud} = 0.97377(11)(15)(19)$, where the first error is the combined experimental ft error and theoretical nuclear structure uncertianty; the second is associated with nuclear coulomb distortion effects; and the last error is the theoretical hadronic structure uncertainty.

New physics can contribute to $\Delta\hat{r}_\beta - \Delta\hat{r}_\mu$ either at tree-level or through loop effects. In SUSY, for example, tree-level effects arise if one allows for terms that violate lepton number[29,30]. Such interactions can be completely supersymmetric and not forbidden by any other symmetry principle[e]. A CKM unitarity deviation

[e]The analogous terms that violate baryon number must be vanishingly small, however, in order to avoid proton decay that is too rapid.

of any sign could be compatible with the presence of such "R parity" violating interactions, so long as one permits effects in both μ and β-decays simultaneously[30]. If one imposes R parity conservation, supersymmetric corrections arise at loop level. My former student A. Kurylov and I analyzed these corrections and found that a unitarity deviation could have interesting implications for the spectrum of supersymmetric particles and models people construct that try to explain why these particles are heavier than the SM particles[31]. From this perspective, then, it is important to know whether or not the CKM matrix is unitarity.

β-decay correlations. Another interesting tool for probing new weak scale physics with β-decay is to study the spectral shape, spatial distribution, and polarization of the outgoing β particles. These characteristics of the spectrum are described by various correlations, as noted many years ago by Jackson, Treiman, and Wyld[32] (for recent discussions, see also [33,34,35]) . For our purposes, it is useful to write the partial rate as

$$d\Gamma \propto \mathcal{N}(E_e)\Bigg\{1 + a\frac{\vec{p}_e \cdot \vec{p}_\nu}{E_e E_\nu} + b\frac{\Gamma m_e}{E_e} + \langle\vec{J}\rangle \cdot \left[A\frac{\vec{p}_e}{E_e} + B\frac{\vec{p}_\nu}{E_\nu} + D\frac{\vec{p}_e \times \vec{p}_\nu}{E_e E_\nu}\right] \tag{3}$$

$$+\vec{\sigma} \cdot \left[N\langle\vec{J}\rangle + G\frac{\vec{p}_e}{E_e} + Q'\hat{p}_e\hat{p}_e \cdot \langle\vec{J}\rangle + R\langle\vec{J}\rangle \times \frac{\vec{p}_e}{E_e}\right]\Bigg\} d\Omega_e d\Omega_\nu dE_e,$$

where $\mathcal{N}(E_e) = p_e E_e (E_0 - E_e)^2$; E_e (E_ν), $\vec{p}_e$ $(\vec{p}_\nu)$, and $\vec{\sigma}$ are the β (neutrino) energy, momentum, and polarization, respectively; $\vec{J}$ is the polarization of the decaying nucleus; and $\Gamma = \sqrt{1-(Z\alpha)^2}$.

The various coefficients of the various correlations in this expression are sensitive to low-energy CC interactions that differ from the $(V-A)\otimes(V-A)$ structure of the SM. It is helpful to characterize these interactions using an effective four fermion Lagragian[36]

$$\mathcal{L}^{\beta-\mathrm{decay}} = -\frac{4G_\mu}{\sqrt{2}} \sum_{\gamma,\,\epsilon,\,\delta} a^\gamma_{\epsilon\delta}\ \bar{e}_\epsilon \Gamma^\gamma \nu_e\ \bar{u}\Gamma_\gamma d_\delta \tag{4}$$

where the sum is over all Dirac matrices $\Gamma^\gamma = 1$ (S), γ^α (V), and $\sigma^{\alpha\beta}/\sqrt{2}$ (T) [f]. and fermion chiralities (ϵ,δ). Within the SM, only a^V_{LL} is non-vanishing, and in this case one has

$$a^V_{LL} = V_{ud}\ (1 + \Delta\hat{r}_\beta - \Delta\hat{r}_\mu) \quad , \tag{5}$$

in accordance with Eq. (2) above.

[f] The normalization of the tensor terms corresponds to the convention adopted in Ref. [37]

Contributions to some of the other $a^{\gamma}_{\epsilon\delta}$ that arise at tree-level in models with RH gauge bosons or leptoquarks have been reviewed extensively by Herczeg[33], and I refer the reader to his work for further details. Here, I wish to summarize a recent analysis of non-$(V-A)\otimes(V-A)$ β-decay operators induced at one-loop order in SUSY[36]. These effects arise when the scalar superpartners of left- and right-handed fermions mix through interactions that break supersymmetry known as triscalar interactions. If $\tilde{f}_{L,R}$ denote these scalar fermions then their mass matrix

$$\mathbf{M}^2_{\tilde{\mathbf{f}}} = \begin{pmatrix} \mathbf{M^2_{LL}} & \mathbf{M^2_{LR}} \\ \mathbf{M^2_{LR}} & \mathbf{M^2_{RR}} \end{pmatrix} \tag{6}$$

will lead to L-R mixing when the off-diagonal entry is non-zero[g]. In the MSSM this term is

$$\mathbf{M^2_{LR}} = \mathbf{M^2_{RL}} = \begin{cases} v\left[\mathbf{a_f}\sin\beta - \mu\mathbf{Y_f}\cos\beta\right] \ , \tilde{u} - \text{type sfermion} \\ v\left[\mathbf{a_f}\cos\beta - \mu\mathbf{Y_f}\sin\beta\right] \ , \tilde{d} - \text{type sfermion} \end{cases} . \tag{7}$$

The MSSM contains two Higgs doublets and $\tan\beta = v_u/v_d$ gives the ratio of the vacuum expectation value of their neutral components, with $v = \sqrt{v_u^2 + v_d^2}$. The quantity μ is a dimensionful parameter that characterizes a supersymmetric coupling between the two Higgs doublets. The matrices $\mathbf{Y_f}$ and $\mathbf{a_f}$ are the 3×3 Yukawa and soft triscalar couplings. The presence of the latter breaks supersymmetry, and in many models, it is often assumed to be proportional to $\mathbf{Y_f}$. Under this "alignment" assumption, the off-diagonal term $\mathbf{M^2_{LR}}$ is suppressed relative to the diagonal terms in the mass matrix for the first and second generation sfermions.

It would be interesting to test alignment assumption for the first two generations, yet doing so with collider studies could be difficult. The study of β-decay correlations, in contrast, may provide a targeted means for carrying out such a test. If the L-R mixing terms are not Yukawa-suppressed and if the mixing is close to maximal, then one-loop effects involving L-R mixed sfermions would induce scalar and tensor operators in $\mathcal{L}^{\beta-\mathrm{decay}}$. The resulting effects in the partial rate would appear in the Fierz interference coefficient, b; the energy-dependent components of the neutrino asymmetry parameter, B, and spin-polarization correlation coefficient Q'; and the energy-independent term in the spin-polarization correlation N. Superallowed nuclear decays, for example, are sensitive to the scalar contribution to b. In the limit that we neglect the chargino ($\tilde{\chi}^{\pm}$) and neu-

[g]In principle, the matrix can also mix superpartners of different generations, so $\mathbf{M^2_{AB}}$ are 3×3 matrices in flavor space.

tralino ($\tilde{\chi}^0$) mixing matrices, we have

$$b_F \approx \frac{2\alpha}{3\pi}\left(\frac{g_S}{g_V}\right)$$
$$\times \mathrm{Re}\, Z_L^{1m} Z_L^{4m*}\left[\left(Z_D^{1i*} Z_D^{4i}\right) M_{\tilde{Z}}^2 \mathcal{F}_1 - \left(Z_U^{1i*} Z_U^{4i}\right) M_{\tilde{Z}}^2 M_{\tilde{\chi}^+} M_{\tilde{\chi}^0} \mathcal{F}_2\right].(8)$$

Here, g_V and g_S are vector and scalar form factors of the nucleon; $\mathcal{F}_{1,2}$ are loop functions of the superpartner masses; and the Z_F^{ij} are elements of the matrices that rotate the sfermion weak eigenstates into corresponding mass eigenstates for superpartners of the charged leptons (L), down (D), and up (U) quarks. The combinations $Z_F^{1m} Z_F^{4m*}$ appearing in Eq. (8) are non-zero only in the presence of L-R mixing among sfermionic superpartners of first generation fermions F.

How precisely would studies of the β spectrum need to be in order to probe these effects at an interesting level? For superpartner masses of order the electroweak scale, the products of $M_{\tilde{Z}}^2$ and the loop functions can be as large as $\mathcal{O}(10^{-1})$; the prefactor $2\alpha/3\pi$ is $\lesssim \mathcal{O}(10^{-2})$; so for large L-R mixing for which the products $Z_F^{1m} Z_F^{4m*}$ are $\mathcal{O}(1)$ the net effect on b_F can approach 10^{-3}. The present limits on b_F are 0.0026(26). With substantial improvements in sensitivity, measurements of the β spectrum that probe this term could search for large L-R mixing and test the alignment hypothesis. Determinations of the correlation coefficients with similar levels of sensitivity would serve the same purpose. Future measurements of the energy-dependence of the neutrino asymmetry parameter B at the few $\times 10^{-4}$ level appear feasible using cold or ultracold neutrons at the Spallation Neutron Source. In either case, null results at this level would point to L-R mixing that is substantially non-maximal and lend experimental credence to the alignment hypothesis.

4. Weak Interactions of Quarks

The weak interactions of quarks at low energies remain enigmatic, despite decades of theoretical and experimental scrutiny. While we have no strong reasons to doubt the the SM prediction for the structure of the elementary weak quark-quark interaction, we have only a limited grasp of the ways in which this interaction becomes manifest in strongly-interacting systems where nonperturbative QCD affects the weak interaction dynamics. In the purely strong interaction sector, the use of symmetries such as chiral, heavy quark, large N_C, and heavy quark symmetries – and the effective theories built on them – have given us powerful tools for explaining strong interaction dynamics without having to compute them from first principles in QCD. Ideally, similar methods would help us explain the hadronic weak interaction (HWI), but our experience is thus far not encouraging. In the $\Delta S = 1$

sector, for example, one cannot simultaneously account for the parity conserving and parity violating (PV) nonleptonic decays of hyperons using chiral symmetry. Similarly, the radiative decays of hyperons produce significantly larger PV asymmetries than one would expect based on the breaking of vector SU(3) symmetry by the strange quark mass. Even the well-known $\Delta I = 1/2$ rule defies explanation based on symmetry considerations.

It may be that our inability to apply successfully the approximate symmetries of QCD to the $\Delta S = 1$ HWI results from the presence of the strange quark with its problematic mass of order the QCD scale. Or, it may be that we are missing a more fundamental aspect of the dynamics involving the interplay of weak and strong interactions of quarks that apply to both light quarks and strange quarks. In order to find out, it is useful to study the $\Delta S = 0$ HWI which is largely devoid of strange quark effects. The difficulty, however, is that we can experimentally access only the PV part of this interaction, since the strong and electromagnetic interactions are considerably larger than the parity conserving $\Delta S = 0$ HWI at low energies.

Historically, the PV $\Delta S = 0$ HWI has been probed using a combination of polarized proton scattering experiments and observation of PV processes in nuclei. The latter have been quite attractive since accidents of nuclear structure can amplify the PV effects making them easier to access experimentally. Unfortunately, the theoretical interpretation of nuclear observables is complicated by nuclear structure, and to date, we have not obtained a consistent description of the $\Delta S = 0$ HWI using nuclear probes. As people at this meeting are well aware, the standard framework for describing this interaction in nuclei has been a meson-exchange model, popularized by Desplanques, Donoghue, and Holstein. The model contains seven PV meson-nucleon couplings, h_M^i, and different nuclear experiments are sensitive to different combinations of these coupling. One such coupling, h_π^1, has received considerable attention, since it is the only PV pion-nucleon coupling and, therefore, the only one that parameterizes the strength of the long-range PV force between nucleons. Measurements of the PV γ-decays of ^{18}F imply that this coupling is consistent with zero and that there is no appreciable, long-range component to the $\Delta S = 0$ HWI. On the other hand, the results for the anapole moment of ^{133}Cs obtained in atomic PV by the Boulder group imply the presence of a large h_π^1. Unless the cesium result is an aberration, there is clearly something about the HWI in nuclei that we do not understand (for a recent discussion of these and other issues, see Ref. [38]).

Fortunately, there exists a way forward. Experimentally, new techniques involving few-body systems make measurements of $\mathcal{O}(10^{-7})$ PV effects in this arena feasible where there were not two decades ago. Theoretically, the impressive developments using Green's function and variational Monte Carlo methods make it possible to perform *ab initio* computations in few-body systems starting from a given nucleon-nucleon potential. In addition, the framework for describing the $\Delta S = 0$ HWI has been reformulated using effective field theory (EFT), thereby allowing one to circumvent the untestable model-dependence of the meson-exchange framework[39]. What has emerged from this confluence of developments is the potential of a new program of few-body hadronic PV studies that could determine the PV "low-energy constants" of the corresponding EFT to lowest order in a way that is free from nuclear structure and hadronic model uncertainties. Completion of this program will require (i) new measurements with neutrons and few-body nuclei at facilities such as the SNS, and (ii) new theoretical computations of the corresponding observables using the EFT framework. I am optimistic that completion of this program will lead to a deeper understanding of the $\Delta S = 0$ HWI and shed new light on the long standing puzzles in the $\Delta S = 1$ sector.

Why is such a program of interest to a future rare isotope facility? The answer is that it could help us better understand the nucleus as a "laboratory" for searching for new physics. At the most basic level, the low-energy HWI involves nuclear and hadronic matrix elements of four quark operators. The program outlined above will help us better understand the dynamics of such matrix elements in the few-nucleon system. Moreover, if the program is successful, we will have in hand a well-determined, PV NN potential to $\mathcal{O}(p)$. This potential could then be used to compute nuclear PV observables, such as the ^{18}F γ-decay polarization or anapole moments of complex nuclei like cesium or francium that could be studied with a rare isotope facility. A comparison of such computations and experimental results will teach us whether or not an EFT weak interaction potential can adequately describe the dynamics of four-quark operators in the nuclear environment – rather than trying to use nuclear observables to determine the potential in a model-dependent way.

In the best of all possible worlds, the EFT approach to treating four-quark operators in nuclei will be successful, thereby allowing us to make progress in the interpretation of another important class of studies where four-quark operators can contribute: neutrinoless double β-decay ($0\nu\beta\beta$). The most important aspect of $0\nu\beta\beta$ is that it may tell us that neutrinos are Majorana fermions. However, people in the business would also like to use it to determine the absolute scale of neutrino mass. This second use of $0\nu\beta\beta$ makes sense so long as there exist no competing contributions to the decay rate from heavy particles with lepton number violating interactions. Unfortunately, reasonable candidates for such particles exist – such as the scalar fermions of SUSY when R parity is not conserved. If the mass scale of such particles is not too different from the electroweak scale, then their contributions to the $0\nu\beta\beta$ rate can be comparable to those from the exchange of a light Majorana neutrino[40]. In this case, one would need to compute the effects of heavy particle exchange in order to separate them out from the possible light Majorana neutrino exchange, and doing so requires calculating nuclear matrix elements of the four-quark operators generated by heavy particle exchange . If our EFT methods for the HWI in nuclei are successful for that problem, then the same techniques could be used with some confidence in the case of heavy particle contributions to $0\nu\beta\beta$, thereby sharpening the interpretation of these important nuclear weak decay experiments. To this end, an EFT formulation for heavy particle contributions to $0\nu\beta\beta$ has recently been developed[41].

5. Conclusions

The primary aim of a new rare isotope facility will, of course, be to to study novel aspects of nuclear structure that are not accessible with present facilities. Carrying out such studies will be an important part of the overall mission of nuclear physics. At the same time, such a facility will provide new opportunities to use nuclei as laboratories to probe the dynamics of quarks at times that preceded the confinement era, to look for evidence of new forces that affected quarks during early times, and to sort out the properties of neutrinos. I hope that in this talk I have conveyed a sense of the opportunities at hand with such a facility – opportunities that I hope our community will one day realize.

Acknowledgments

I wish to thank the organizers of the workshop for financial support and hospitality. This work was supported in part under U.S. DOE contract DE-FG02-05ER41361 and NSF grant PHY-0555674.

References

1. *Report to the Nuclear Science Advisory Committee: Guidance for Implementing the 2002 Long Range Plan*, R. Tribble *et al* (2005): http://www.sc.doe.gov/np/nsac/nsac.htm
2. M. J. Ramsey-Musolf, "Probing the fundamental symmetries of the early universe: The low energy arXiv:hep-ph/0603023.
3. J. Erler and M. J. Ramsey-Musolf, Prog. Part. Nucl. Phys. **54**, 351 (2005) [arXiv:hep-ph/0404291].
4. A. D. Sakharov, Pisma Zh. Eksp. Teor. Fiz. **5**, 32 (1967) [JETP Lett. **5**, 24 (1967)].
5. S. Eidelman *et al.* [Particle Data Group Collaboration], Phys. Lett. B **592**, 1 (2004).
6. D. N. Spergel *et al.* [WMAP Collaboration], Astrophys. J. Suppl. **148**, 175 (2003) [arXiv:astro-ph/0302209].

7. M. E. Shaposhnikov, Nucl. Phys. B **287**, 757 (1987).
8. G. R. Farrar and M. E. Shaposhnikov, Phys. Rev. Lett. **70**, 2833 (1993) [Erratum-ibid. **71**, 210 (1993)] [arXiv:hep-ph/9305274].
9. A. Riotto and M. Trodden, Ann. Rev. Nucl. Part. Sci. **49**, 35 (1999) [arXiv:hep-ph/9901362].
10. A. Riotto, Phys. Rev. D **58**, 095009 (1998) [arXiv:hep-ph/9803357]; M. Carena, J. M. Moreno, M. Quiros, M. Seco and C. E. M. Wagner, Nucl. Phys. B **599**, 158 (2001) [arXiv:hep-ph/0011055]; M. Carena, M. Quiros, M. Seco and C. E. M. Wagner, Nucl. Phys. B **650**, 24 (2003) [arXiv:hep-ph/0208043]; T. Konstandin, T. Prokopec and M. G. Schmidt, Nucl. Phys. B **716**, 373 (2005) [arXiv:hep-ph/0410135]. C. Lee, V. Cirigliano and M. J. Ramsey-Musolf, Phys. Rev. D **71**, 075010 (2005) [arXiv:hep-ph/0412354]; T. Konstandin, T. Prokopec, M. G. Schmidt and M. Seco, arXiv:hep-ph/0505103.
11. V. Cirigliano, S. Profumo and M. J. Ramsey-Musolf, JHEP **0607**, 002 (2006) [arXiv:hep-ph/0603246].
12. M. Pospelov and A. Ritz, Annals Phys. **318**, 119 (2005) [arXiv:hep-ph/0504231].
13. C. S. Wu, E. Ambler, R. W. Hayward, D. D. Hoppes and R. P. Hudson, Phys. Rev. **105**, 1413 (1957).
14. R. L. Garwin, L. M. Lederman and M. Weinrich, Phys. Rev. **105**, 1415 (1957).
15. I. S. Towner and J. C. Hardy, Phys. Rev. C **72**, 055501 (2005) [arXiv:nucl-th/0508039].
16. E. Blucher *et al.*, arXiv:hep-ph/0512039.
17. P. Post and K. Schilcher, Eur. Phys. J. C **25**, 427 (2002) [arXiv:hep-ph/0112352].
18. J. Bijnens and P. Talavera, Nucl. Phys. B **669**, 341 (2003) [arXiv:hep-ph/0303103].
19. V. Cirigliano, G. Ecker, M. Eidemuller, R. Kaiser, A. Pich and J. Portoles, JHEP **0504**, 006 (2005) [arXiv:hep-ph/0503108].
20. D. Becirevic *et al.*, Nucl. Phys. B **705**, 339 (2005) [arXiv:hep-ph/0403217].
21. M. Okamoto [Fermilab Lattice Collaboration], arXiv:hep-lat/0412044.
22. N. Tsutsui *et al.* [JLQCD Collaboration], PoS **LAT2005**, 357 (2005) [arXiv:hep-lat/0510068].
23. C. Dawson, T. Izubuchi, T. Kaneko, S. Sasaki and A. Soni, PoS **LAT2005**, 337 (2005) [arXiv:hep-lat/0510018].
24. W. J. Marciano and A. Sirlin, Phys. Rev. Lett. **96**, 032002 (2006) [arXiv:hep-ph/0510099].
25. J. C. Hardy, I. S. Towner and G. Savard, arXiv:nucl-ex/0511051.
26. G. Savard *et al.*, Phys. Rev. Lett. **95**, 102501 (2005).
27. A. Serebrov *et al.*, Phys. Lett. B **605**, 72 (2005) [arXiv:nucl-ex/0408009].
28. H. Abele *et al.*, Phys. Rev. Lett. **88**, 211801 (2002) [arXiv:hep-ex/0206058].
29. V. D. Barger, G. F. Giudice and T. Han, Phys. Rev. D **40**, 2987 (1989).
30. M. J. Ramsey-Musolf, Phys. Rev. D **62**, 056009 (2000) [arXiv:hep-ph/0004062].
31. A. Kurylov and M. J. Ramsey-Musolf, Phys. Rev. Lett. **88**, 071804 (2002) [arXiv:hep-ph/0109222].
32. J. D Jackson, S. B. Treiman, and H. W. Wyld, Jr., Phys. Rev. **106**, 517 (1957); Nucle. Phys. **4**, 206 (1957).
33. P. Herczeg, Prog. Part. Nucl. Phys. **46**, 413 (2001).
34. J. Deutsch and P. Quin in *Precision Tests of the Standard Electroweak Model*, P. Langacker, ed., World Scientific (Singapore) 1995, p. 706.

35. N. Severijns, M. Beck and O. Naviliat-Cuncic, arXiv:nucl-ex/0605029.
36. S. Profumo, M. J. Ramsey-Musolf and S. Tulin, arXiv:hep-ph/0608064.
37. F. Scheck, *Electroweak and Strong Interactions: An Introduction to Theoretical Particle Physics*, Springer Verlag, 1996, p.282.
38. M. J. Ramsey-Musolf and S. A. Page, arXiv:hep-ph/0601127.
39. S. L. Zhu, C. M. Maekawa, B. R. Holstein, M. J. Ramsey-Musolf and U. van Kolck, Nucl. Phys. A **748**, 435 (2005) [arXiv:nucl-th/0407087].
40. V. Cirigliano, A. Kurylov, M. J. Ramsey-Musolf and P. Vogel, Phys. Rev. Lett. **93**, 231802 (2004) [arXiv:hep-ph/0406199].
41. G. Prezeau, M. Ramsey-Musolf and P. Vogel, Phys. Rev. D **68**, 034016 (2003) [arXiv:hep-ph/0303205].

NUCLEAR EFFECTIVE FIELD THEORIES*

U. VAN KOLCK
Department of Physics, University of Arizona
Tucson, AZ 85721, USA
E-mail: vankolck@physics.arizona.edu

I review various effective field theories that have been developed to understand nuclear structure and reactions, with emphasis on some recent developments.

1. Effective Field Theories

Despite an array of interesting people in this workshop, I deeply missed the vibrant presence of Vijay Pandharipande, who passed away earlier this year. Although (or, perhaps, *because*) Vijay and I would almost always end up arguing about some important point, he was a constant inspiration to me, particularly for the way he carried out his research program. Here I am going to relate some of the recent and not-so-recent developments in a research program parallel to Vijay's, that of developing nuclear effective field theories (EFTs). I cannot be complete in my review. Instead, I am going to concentrate on some issues of relevance to modern theories of nuclear structure and reactions, and indirectly to a RIA-like machine. Among other topics, I will mention work done with Vijay on the role of delta isobars.

Nuclear structure involves energies that are much smaller than the typical QCD mass scale, $M_{QCD} \sim 1$ GeV. This is a common situation in physics: an "underlying" theory is valid at a mass scale $M_{\rm hi}$, but we want to describe processes at momenta Q of the order of a lower scale $M_{\rm lo} \ll M_{\rm hi}$. What are the "effective" interactions that will ensure that low-energy observables are described correctly? Understanding effective interactions in the many-body problem has long been recognized as important by the nuclear-structure community. Yet, it does not seem widely appreciated within this community that the same is true for *any* problem (with the possible exception of a final theory of everything, if such a thing exists), including the few-nucleon forces that are a fundamental input to

*This work is supported by the U.S. Department of Energy and the Sloan Foundation.

the many-nucleon problem. EFT provides a systematic framework to construct effective nuclear interactions.

EFT starts with the observation that effective interactions consist of the sum of *all* possible interaction terms in a Lagrangian that involves only the fields representing low-energy degrees of freedom. Because of the uncertainty principle, each of these interaction terms can be taken as a local combination of derivatives of the fields. If the "integrating out" of the high-energy degrees of freedom is done appropriately, the effective Lagrangian will have the same symmetries as the underlying theory. The details of the underlying dynamics, on the other hand, are contained in the interaction strengths. The latter also depend on the details of how the low- and high-energy degrees of freedom are separated. This separation requires the introduction of a cutoff parameter Λ with dimensions of energy. Both the quantum effects represented by loops *and* the interaction strengths depend on Λ (the interaction strengths "run"). However, the cutoff procedure is arbitrary, so by construction observables are independent of Λ ("renormalization-group invariance"). The T matrix for any low-energy process acquires the schematic form

$$T(Q \sim M_{\rm lo}) = N \sum_{\nu=\nu_{\rm min}}^{\infty} c_\nu(M_{\rm hi}, \Lambda) \left(\frac{Q}{M_{\rm hi}}\right)^{\nu} F_\nu \left(\frac{Q}{M_{\rm lo}}; \frac{\Lambda}{M_{\rm lo}}\right), \qquad (1)$$

where N is a common normalization factor, ν is a counting index starting at some value $\nu_{\rm min}$, the c_ν's are parameters, and the F_ν's are calculable functions. We must have

$$\frac{\partial T(Q \sim M_{\rm lo})}{\partial \Lambda} = 0. \qquad (2)$$

In order to maintain predictive power in the EFT it is necessary to truncate the sum above in such a way that the resulting Λ dependence can be decreased systematically with increasing order. We call such ordering "power counting". When we do not know or cannot solve the underlying theory, the minimal assumption is that renormalized interactions are natural, that is, that their sizes are of the order of magnitude of the most important contributions to their running. This guess is confirmed *a posteriori*, by checking renormalization-group invariance and convergence of the truncation after the data is fitted order by order.

For the last ten years or so we have been developing EFTs for systems of few nucleons [1]. The goal is not only to understand traditional nuclear physics from a Standard Model standpoint, but also to develop a systematic framework that will allow to quantify deviations from the Standard Model. (I will omit the latter aspect here, though, because Mike Ramsey-Musolf has already described some of this work in his talk [2].) EFTs have been constructed for various relevant $M_{\rm lo}$'s. I deal with them in turn.

2. Pionful EFT

Much of the work on nuclear EFTs has been devoted to the EFT where $M_{\rm hi} \sim M_{QCD}$ and $M_{\rm lo} \sim m_\pi$, the pion mass. In the sector of $A = 0, 1$ nucleons this EFT reduces to well-understood chiral perturbation theory (ChPT) (for a review, see Ref. [3]), and for $A \geq 2$ fairly sophisticated potentials have been constructed (see the reviews [1]). Yet, as we are going to see, recent developments [4,5] show that these potentials are incomplete.

In this "pionful" EFT the degrees of freedom are nucleons, pions, and delta isobars; others, such as the ρ meson of mass $m_\rho \sim M_{QCD}$, can be integrated out and accounted for indirectly through a "multipole" expansion in powers of Q/m_ρ, *etc.* QCD has (an apparently) exact invariance under Lorentz transformations, and approximate invariance under parity, time-reversal, and chiral symmetries. Lorentz invariance can be easily incorporated for nucleons and deltas in a non-relativistic Q/m_N expansion. Chiral symmetry imposes strong constraints on pion interactions. For two quark flavors (for simplicity), $SU(2)_L \times SU(2)_R$ is an approximate invariance of the dynamics, but the vacuum is symmetric only under the diagonal $SU(2)_{L+R}$ subgroup. Pions can be viewed as pseudo-Goldstone bosons living on a "chiral circle" of radius $f_\pi \simeq 92$ MeV, the pion decay constant.

The techniques to build the most general Lagrangian involving pion (π) and fermion (Ψ) fields are by now standard —see, for example, Refs. [3,1]. Schematically, the Lagrangian can be written as

$$L_\pi = \sum_{d,p,f} c_{d,p,f} \left(\frac{\partial, m_\pi, m_\Delta - m_N}{M_{QCD}} \right)^d \left(\frac{\pi}{f_\pi} \right)^p \left(\frac{\Psi^\dagger \Psi}{M_{QCD} f_\pi^2} \right)^{f/2} f_\pi^2 M_{QCD}^2. \tag{3}$$

One can show that chiral symmetry implies that the "chiral index" $\Delta \equiv d + f/2 - 2 \geq 0$. Here, the coefficients of the operators were written in terms of dimensionless parameters $c_{d,p,f}$'s. The crucial issue for power counting is to estimate the size of these parameters.

For $A = 0, 1$ processes, because pions and photons typically carry energies of $O(Q)$, an intermediate state differs in energy from the initial state by an energy $O(Q)$. As a consequence, each additional loop is found [6] to be suppressed by a power of $(Q/4\pi f_\pi)^2$, and naturalness implies that the $c_{d,p,f=0,2}$'s are of $O(1)$. Amplitudes can be written in the form (1) with $\nu = 2 - A + 2L + \sum_i V_i \Delta_i \geq 2 - A \equiv \nu_{min}$, where in a Feynman diagram L is the number of loops and V_i the number of vertices with index Δ_i. Amplitudes can therefore be calculated in perturbation theory in Q/M_{QCD}.

In the $A \geq 2$ sector, on the other hand, power counting is much more subtle. Weinberg [7] recognized that there is a breakdown of ChPT power counting due to "reducible" intermediate states that differ in energy from initial states only by the kinetic energy of nucleons. The resulting small energy denominators of $O(Q^2/m_N)$ represent an infrared enhancement in amplitudes of $O(m_N/Q)$ relative to ChPT. Since one-pion exchange (OPE) between nucleons is $O(1/f_\pi^2) \equiv O(4\pi/m_N M_{nuc})$ and an intermediate state contributes $O(im_N Q/4\pi)$ to the two-nucleon ($2N$) T matrix, an iteration of OPE gives an extra $O(iQ/M_{nuc})$. The series is non-perturbative, and thus can support a bound state, at a momentum $M_{nuc} = 4\pi f_\pi^2/m_N \sim 100$ MeV, or energy $Q^2/m_N \sim M_{nuc}^2/M_{QCD} \sim 10$ MeV.

While pions can give rise to nuclei of roughly the observed size, a model-independent, quantitative description requires also short-range interactions. One expects diagrams with "irreducible" intermediate states, which do not contain the infrared enhancement, to contribute to the running of the short-range parameters like in ChPT, leading to $c_{d,p,f\geq 4}$'s of $O(1)$. Under this assumption, Weinberg [7] suggested that a generic nuclear amplitude should be obtained from a potential calculated as the sum of irreducible sub-diagrams ordered according to standard ChPT power counting, followed by iteration of the potential to all orders via the Lippmann-Schwinger or Schrödinger (S) equations. In leading order (LO) in Weinberg's power counting, the nuclear potential is a sum of $2N$ potentials given by OPE and two non-derivative, chirally invariant delta-function interactions that contribute in the S channels. In subleading orders, other components of the nuclear potential are present such as contact interactions with derivatives or powers of m_π^2, two- (TPE) and more-pion exchanges, and three- ($3N$) and more-nucleon interactions. While this potential shares many similarities with conventional ones, it also has some novel features. Of special relevance are TPE components with the correct chiral symmetry properties. In the $2N$ sector, TPE [8] resembles a van der Waals force and has been "seen" in the Nijmegen partial-wave analysis (PWA) [9]. In the $3N$ sector, the TPE component [10] corrects the well-known Tucson-Melbourne force [11].

However, we now know that Weinberg's power counting is *in*consistent. Following an earlier claim based on a perturbative argument [12], it has been shown [4] that reducible loops lead to new contributions to the non-perturbative running of some of the $c_{d,p,f=4}$'s, which are then larger than assumed in Weinberg's power counting. OPE is a singular potential, which is attractive in several partial waves. In a given partial wave, both solutions of the S equation with an attractive singular potential oscillate at small distances, and a phase remains that is *not* fixed by the long-range potential. This phase determines low-energy observables but it can only be made cutoff independent by a short-range interaction [13]. The corresponding interaction is absent [4] in Weinberg's LO *i)* in the $^1S_0$ channel, where the

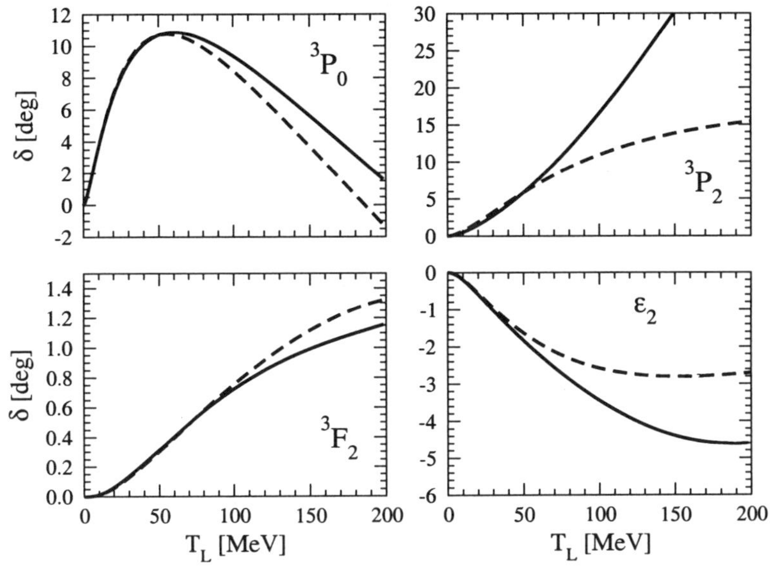

Figure 1. Attractive $2N$ triplet phase shifts (in degrees) as function of the laboratory energy (in MeV): EFT in LO (solid line) and Nijmegen PWA [15] (dashed line). From Ref. [4].

interference between delta function and OPE demands a chiral-symmetry breaking counterterm; and *ii)* in triplet channels where the tensor force is attractive, which requires chiral-invariant counterterms also in channels other than $^3S_1$-$^3D_1$. The short-range parameters needed to renormalize iterated OPE are enhanced in the infrared and driven by pion parameters, effects of derivatives scaling roughly as Q/lf_π (for large angular momentum l), rather than as Q/M_{QCD} if the renormalization were driven by loops in the potential. The factor l accounts for the centrifugal barrier that ensures that at sufficiently large l OPE can be treated in perturbation theory [14] at $Q \sim m_\pi$, and that the corresponding counterterms scale as assumed in Weinberg's power counting. In contrast, for a finite number of low partial waves resummation is necessary and the cutoff dependence can be absorbed by one promoted counterterm per partial wave, which then exhibits a limit-cycle-like behavior. The fitted LO $2N$ phase shifts [4] come out well at low energies, as illustrated in Fig. 1. No residual cutoff dependence is observed for the triton binding energy [4], suggesting that the $c_{d,p,f\geq 6}$'s might indeed be of $O(1)$ as assumed by Weinberg.

What about subleading orders? Corrections up to $O((Q/M_{QCD})^3)$ can be treated in first-order perturbation theory in the background of the LO potential (distorted-wave Born approximation). The best strategy is *not* to integrate out delta isobars [5]. The EFT folklore is that, because typical nuclear energies are $\sim M_{nuc}^2/m_N \ll m_\Delta - m_N$, one can integrate out the delta with very small error. The problem, however, is that the short-range interactions that subsume delta contributions need to be obtained from pion-nucleon scattering, where the energy is $\sim m_\pi$, so an error of $O(m_\pi/(m_\Delta - m_N))$, numerically $\sim 1/2$, contaminates the nuclear calculation. Examples are the TPE $2N$ force [8] and the Fujita-Miyazawa $3N$ force [11], where integrating out the delta leads to poor convergence [9] and a 30% overestimate [5], respectively. While these effects can be compensated by higher-order terms, it is easier to simply include the delta explicitly in the corresponding subleading potentials, as done originally [8,10].

Because it includes chiral symmetry breaking explicitly, the pionful EFT is unique in its ability to handle varying pion masses, as required to make the connection with lattice QCD simulations at larger masses. Even though the need for an explicit delta only becomes more acute in this case, it has already been verified without it that $2N$ bound-state momenta acquire more natural values when one varies the pion mass [4,16].

3. Pionless EFT

For the observed value of the quark masses, on the other hand, the bound states have binding momenta somewhat smaller than the estimate $M_{nuc} \sim 100$ MeV: $\aleph_1 \sim \sqrt{m_N B_d} \simeq 45$ MeV for the deuteron and $\aleph_0 \sim \sqrt{m_N B_d'} \simeq 8$ MeV for the $^1S_0$ virtual state. This indicates an amount of fine-tuning, and to describe reactions on these bound states involving momenta $Q \ll m_\pi$ the "pionful" EFT is an overkill. One can instead consider a "pionless" EFT where even the pion cloud is represented by a multipole expansion: $M_{\rm hi} \sim m_\pi \sim M_{nuc}$ and $M_{\rm lo} \sim \aleph$ (with $\aleph$ some average of the $\aleph_i$).

The most general Lagrangian containing nucleon (N) fields and subject to Lorentz and (approximate) parity and time-reversal symmetries has only contact interactions, and can be written in a schematic form similar to (3),

$$L_{no\pi} = \sum_{d,f} c_{d,f} \left(\frac{\partial}{M_{\rm hi}}\right)^d \left(\frac{4\pi N^\dagger N}{m_N M_{\rm hi}^2}\right)^{f/2} \frac{m_N M_{\rm hi}^3}{4\pi}. \tag{4}$$

A more convenient way [1] to write the Lagrangian employs two auxiliary, "dimeron" fields corresponding to the bound states [17]. The final results for observables are, of course, independent of the choice of fields.

As in the pionful EFT, the crucial issue is the size of the $c_{d,f}$'s. If there were no fine-tuning, one would expect them to be of $O(1)$. In this case, amplitudes would, like in ChPT, be perturbative, and bound states would be outside the range of validity of the theory. The $c_{d,f}$'s have, therefore, to contain powers of $\aleph^{-1}$. In fact, it can be shown [18] that if the $c_{d,f=4}$'s that contribute to S waves are $O((M_{\rm hi}/\aleph)^{1+d/2})$, then in the $2N$ system the leading, non-derivative interactions have to be summed to all orders —giving rise to the observed shallow S-wave bound states— while higher-derivative interactions give perturbative corrections suppressed by powers of $Q/M_{\rm hi}$. The $2N$ T matrix has the form (1) with $N = 4\pi/m_N M_{\rm hi}$ and $\nu = \sum_i V_i d_i - p + L$, where in a diagram V_i is the number of vertices with d_i derivatives, p is the number of dimeron propagators, and L is the number of loops. It is equivalent order by order to that of the effective-range expansion. The resulting phase shifts [18,1] converge to empirical values for $Q \lesssim m_\pi$. The deuteron binding energy is found to be $B_d = 1.91$ MeV in next-to-leading order (NLO), to be compared with the experimental value of 2.22 MeV. In addition, many low-energy reactions involving the deuteron have been studied, as reviewed in Ref. [1].

The $3N$ system is more interesting. In all but the $S_{1/2}$ wave, $3N$ forces appear only at high orders, and very precise results [19] for nucleon-deuteron (Nd) scattering follow with parameters fully determined from $2N$ scattering. For example, excellent agreement with data is achieved in the $S_{3/2}$ channel already at next-to-next-to-leading order (NNLO), the scattering length being postdicted as $a_{3/2} = 6.33 \pm 0.10$ fm, to be compared to the experimental value, 6.35 ± 0.02 fm. One can, thus, do QED-quality nuclear physics with EFT. In the $S_{1/2}$ wave, renormalization-group invariance can only be achieved if the $3N$ contact interactions are also enhanced [20] by powers of $\aleph^{-1}$, so the $c_{d,f=6}$'s are $O((M_{\rm hi}/\aleph)^{2+d/2})$. In this channel, a single non-derivative $3N$ interaction appears in LO, and higher-derivative interactions are smaller by powers of $Q/M_{\rm hi}$. To NLO there is only one parameter not fixed by $2N$ observables; it can be fixed by, say, the $S_{1/2}$ Nd scattering length. As a function of the cutoff it displays an unusual, limit-cycle behavior [20]. The resulting energy dependence of $S_{1/2}$ Nd scattering —see Fig. 2— comes out well. Likewise, the triton binding energy is found to be $B_t = 8.31$ MeV in NLO, to be compared with the experimental value of 8.48 MeV.

The smaller $\aleph$, the closer the $3N$ system is to a limit cycle, where there is a discrete scale invariance manifest in a set of low-energy bound states. It has, in fact, been conjectured [24] that a small change in quark masses would make $\aleph = 0$, that is, that QCD is near an infrared limit cycle.

The pionless EFT simplifies the treatment of light nuclei, but the application to larger nuclei still faces computational challenges. For a first attack on the four-nucleon system, see Ref. [25]. To go beyond that, we are currently implementing the EFT in the No-Core Shell Model (NCSM) [26]. Hopefully, shell structure will

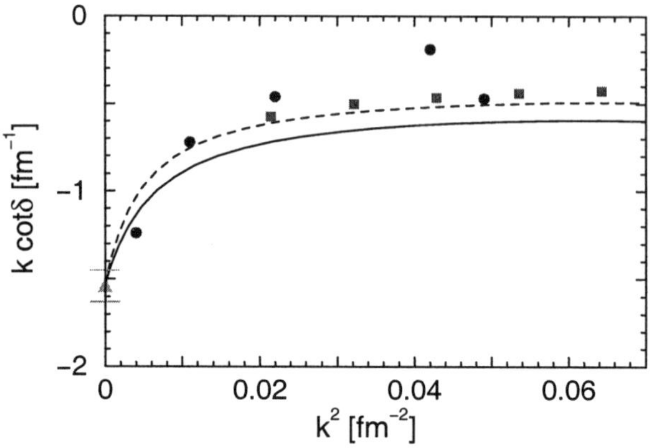

Figure 2. The $S_{1/2}$ Nd K^{-1} matrix (in fm^{-1}) as function of the square of the center-of-mass momentum (in fm^{-2}): EFT results at LO (solid line) and NLO (dashed line), phenomenological potential model [21] (squares), phase-shift analysis [22] (dots), and cold-neutron measurement [23] (triangle). From Ref. [20], courtesy of H. Hammer.

arise in this context.

4. Halos

Another approach for describing larger nuclei is to develop further EFTs, tailored to these systems. As a first step, we can specialize to anomalously-shallow ("halo") nuclei and some reactions of astrophysical interest. A halo system can be defined as one that contains two momentum scales: $M_{\rm hi} \sim \sqrt{m_N E_{\rm core}}$, associated with the excitation energy $E_{\rm core}$ of a tight cluster of nucleons ("core"); and $M_{\rm lo} \sim \sqrt{m_N E_{\rm halo}}$, associated with the energy $E_{\rm halo}$ for the attachment or removal of one or more (halo) nucleons. Such a system exhibits shallow S-matrix poles, either on the imaginary axis (bound states) or elsewhere in the complex momentum plane (resonances). The deuteron and the triton are two- and three-body halo systems, where the core is a single nucleon ($E_{\rm core} \sim m_\pi^2/m_N$ and $E_{\rm halo} \sim B_d$ and B_t). The next-simplest examples involve a ^{4}He core, for which $E_{\rm core} \simeq 20$ MeV, and $E_{\rm halo} \sim 1$ MeV is the energy of the $P_{3/2}$ resonance in nucleon-alpha ($N\alpha$) scattering and the two-neutron removal energy in ^{6}He.

It is natural to generalize the EFT to describe shallow two-body resonances [27], as a step before tackling three-body halo nuclei. The method parallels the one discussed in the previous section for shallow bound states, but with important differences. The relevant degrees of freedom are nucleons and cores, and the symmetries, the same of the pionless EFT. The most convenient way to write the most general Lagrangian is to introduce a dimeron field for the resonance. A

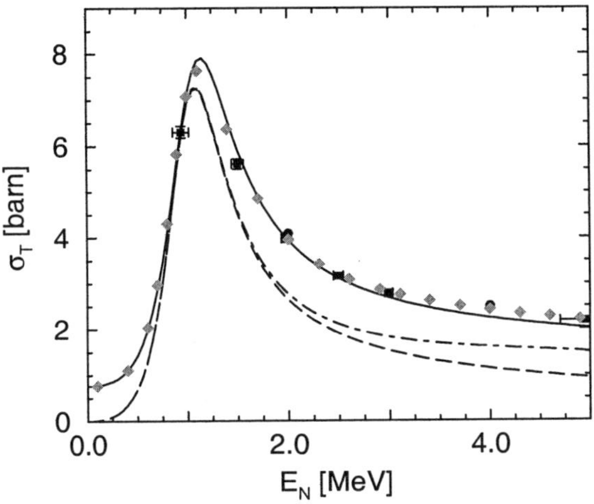

Figure 3. Total $n\alpha$ cross section (in barns) as a function of the neutron energy (in MeV) in the alpha-particle rest frame: EFT in LO (dashed line) and NLO (solid line), evaluated data [28] (diamonds), and experimental data [29] (squares). (The dash-dotted line represents an alternative calculation with a modified power counting where the $P_{1/2}$ resonance is considered shallow.) From Ref. [27].

shallow resonance at $Q \sim M_{\rm lo}$ can only be accounted for if the EFT parameters depend on $M_{\rm lo}$ in a specific way. One finds [27] that generically the nucleon-core T matrix is perturbative, having the form (1) with $N = 2\pi/\mu M_{\rm hi}$ (with μ the reduced mass) and $\nu = \sum_i V_i d_i - 2p + L$. However, in a momentum region of $O(M_{\rm lo}^2/M_{\rm hi})$ around the resonance there is a cancellation in the denominator of the dimeron propagator, and a resummation of the dimeron self-energy is necessary. In this region, the nucleon-core T matrix still has the form (1) but now $\nu = \sum_i V_i d_i - 3p + L$. The corresponding total neutron-alpha ($n\alpha$) cross section [27] is compared with data in Fig. 3, where the success of this resummed description is evident throughout the low-energy region.

The next step is to use the $N\alpha$ interactions determined from $N\alpha$ scattering [27] and the $2N$ interactions determined from $2N$ scattering [18] to calculate the Borromean halo ^{6}He as a ^{4}He $+ n + n$ system [30], the same way we successfully described triton as a $p + n + n$ system [20]. But clearly the theory can be applied to any halo nucleus, including those involved in reactions of astrophysical interest such as $p +\, ^7$Be $\rightarrow\, ^8$B $+ \gamma$, which can be analyzed as was $p + n \rightarrow d + \gamma$ [18]. An important ingredient is the Coulomb interaction, and as a test case we are studying $p\alpha$ scattering [31].

5. Outlook

One of the fundamental goals of nuclear physics is to understand how nuclei arise from QCD. EFT is a central tool in this task. We see explicitly via lattice simulations that from QCD hadronic degrees of freedom emerge subject to the same symmetries as at high energies (except that color gauge invariance is realized trivially, and isospin is enhanced). Because the EFT is the most general dynamics consistent with these symmetries, eventually, EFT parameters will be calculated in terms of QCD parameters (Λ_{QCD} and quark masses).

New phenomena then emerge within the EFT. Halo-like systems (including the deuteron and the triton) arise from fine-tuning. Presumably, the fine-tuning scale will be visible in the calculated parameters, but a precise description will still benefit from fitting the parameters to a set of data. In any case, nuclei can be quite large, and it would be a waste of resources to attempt to obtain these bound states directly from simulations. Rather, a tower of EFTs can be used to capture —with methods, such as the NCSM, developed by the many-body community— the long-distance effects that govern the structure and reactions of few- and (eventually) many-nucleon systems. Limit cycles and shells are examples of the emergent nuclear phenomena that are not evident in QCD.

In this context, what might be the role of a RIA-like machine? The keyword is, I believe, fine-tuning. By studying systems that exist as consequence of fine-tuning, we are amplifying its effects. For the pionful EFT, the case is similar to phenomenological potentials: for example, we expect that the precise form of $3N$ (and perhaps $4N$) forces —and the related role of the delta isobar, a key ingredient of the $3N$ force— will be clarified in systems of extreme isospin. Regarding the pionless and halo EFTs, data will become available to (better) constrain fine-tuned parameters. More generally, it is possible that extreme conditions might guide the development of EFTs for heavier nuclei. Of course, a RIA-like machine will also help with applications such as symmetry tests and reactions of astrophysical interest.

In conclusion, there have been many successes in the EFT program, but still much more to do, in particular, grow to larger nuclei. There is real hope that within the next few years we will witness the description of nuclear structure and reactions in a framework consistent with the Standard Model.

Acknowledgments

I am grateful to my collaborators for the fun work, and to the organizers for the invitation to a timely workshop. This research is supported in part by the U.S. Department of Energy and by the Alfred P. Sloan Foundation.

References

1. U. van Kolck, *Prog. Part. Nucl. Phys.* **43**, 337 (1999); P.F. Bedaque and U. van Kolck, *Ann. Rev. Nucl. Part. Sci.* **52**, 339 (2002); E. Epelbaum, `nucl-th/0509032`.
2. M. Ramsey-Musolf, talk at this workshop.
3. V. Bernard, N. Kaiser, and U.-G. Meißner, *Int. J. Mod. Phys.* **E4**, 193 (1995).
4. S.R. Beane, P.F. Bedaque, M.J. Savage, and U. van Kolck, *Nucl. Phys.* **A700**, 377 (2002); A. Nogga, R.G.E. Timmermans, and U. van Kolck, *Phys. Rev.* **C72**, 054006 (2005).
5. V.R. Pandharipande, D.R. Phillips, and U. van Kolck, *Phys. Rev.* **C71**, 064002 (2005).
6. S. Weinberg, *Physica* **A96**, 327 (1979); A. Manohar and H. Georgi, *Nucl. Phys.* **B234**, 189 (1984).
7. S. Weinberg, *Phys. Lett.* **B251**, 288 (1990); **B295**, 114 (1992).
8. C. Ordóñez, L. Ray, and U. van Kolck, *Phys. Rev.* **C53**, 2086 (1996).
9. M.C.M. Rentmeester, R.G.E. Timmermans, and J.J. de Swart, *Phys. Rev.* **C67**, 044001 (2003).
10. U. van Kolck, *Phys. Rev.* **C49**, 2932 (1994); J.L. Friar, D. Hüber, and U. van Kolck, *Phys. Rev.* **C59**, 53 (1999).
11. J.-I. Fujita and H. Miyazawa, *Prog. Theor. Phys.* **17**, 360 (1957); S.A. Coon, M.D. Scadron, P.C. McNamee, B.R. Barrett, D.W.E. Blatt, and B.H.J. McKellar, *Nucl. Phys.* **A317**, 242 (1979).
12. D.B. Kaplan, M.J. Savage, and M.B. Wise, *Nucl. Phys.* **B534**, 329 (1998).
13. S.R. Beane, P.F. Bedaque, L. Childress, A. Kryjevski, J. McGuire, and U. van Kolck, *Phys. Rev.* **A64**, 042103 (2001).
14. M.C. Birse, `nucl-th/0507077`.
15. V.G.J. Stoks, R.A.M. Klomp, M.C.M. Rentmeester, and J.J. de Swart, *Phys. Rev.* **C48**, 792 (1993).
16. S.R. Beane and M.J. Savage, *Nucl. Phys.* **A713**, 148 (2003); **A717**, 91 (2003). E. Epelbaum, U.-G. Meißner, and W. Glöckle, *Nucl. Phys.* **A714**, 535 (2003).
17. D.B. Kaplan, *Nucl. Phys.* **B494**, 471 (1997).
18. U. van Kolck, `hep-ph/9711222`, in *Proceedings of the Workshop on Chiral Dynamics 1997, Theory and Experiment*, ed. A. Bernstein, D. Drechsel, and T. Walcher (Springer-Verlag, Berlin, 1998); *Nucl. Phys.* **A645**, 273 (1999); D.B. Kaplan, M.J. Savage, and M.B. Wise, *Phys. Lett.* **B424**, 390 (1998); J. Gegelia, `nucl-th/9802038`; J.-W. Chen, G. Rupak, and M.J. Savage, *Nucl. Phys.* **A653**, 386 (1999).
19. P.F. Bedaque and U. van Kolck, *Phys. Lett.* **B428**, 221 (1998); P.F. Bedaque, H.-W. Hammer, and U. van Kolck, *Phys. Rev.* **C58**, R641 (1998); F. Gabbiani, P.F. Bedaque, and H.W. Grießhammer, *Nucl. Phys.* **A675**, 601 (2000).
20. P.F. Bedaque, H.-W. Hammer, and U. van Kolck, *Nucl. Phys.* **A676**, 357 (2000); H.-W. Hammer and T. Mehen, *Phys. Lett.* **B516**, 353 (2001); P.F. Bedaque, G. Rupak, H.W. Grießhammer, and H.-W. Hammer, *Nucl. Phys.* **A714**, 589 (2003).
21. A. Kievsky, S. Rosati, W. Tornow, and M. Viviani, *Nucl. Phys.* **A607**, 402 (1996).
22. W.T.H. van Oers and J.D. Seagrave, *Phys. Lett.* **B24**, 562 (1967); A.C. Philips and G. Barton, *Phys. Lett.* **B28**, 378 (1969).
23. W. Dilg, L. Koester, and W. Nistler, *Phys. Lett.* **B36**, 208 (1971).
24. E. Braaten and H.-W. Hammer, *Phys. Rev. Lett.* **91**, 102002 (2003); E. Epelbaum,

H.-W. Hammer, U.-G. Meißner, and A. Nogga, hep-ph/0602225.
25. L. Platter, H.-W. Hammer, and U.-G. Meißner, *Phys. Lett.* **B607**, 254 (2005).
26. I. Stetcu, B.R. Barrett, and U. van Kolck, in progress.
27. C.A. Bertulani, H.-W. Hammer, and U. van Kolck, *Nucl. Phys.* **A712**, 37 (2002); P.F. Bedaque, H.-W. Hammer, and U. van Kolck, *Phys. Lett.* **B569**, 159 (2003).
28. *Evaluated Nuclear Data Files*, National Nuclear Data Center, Brookhaven National Laboratory (http://www.nndc.bnl.gov/).
29. B. Haesner *et al.*, *Phys. Rev.* **C28**, 995 (1983); M.E. Battat *et al.*, *Nucl. Phys.* **12**, 291 (1959).
30. H.-W. Hammer and U. van Kolck, in progress.
31. R. Higa, C.A. Bertulani, and U. van Kolck, in progress.

PARTICIPANTS OF
THE THIRD ANL/MSU/JINA/INT RIA THEORY MEETING
ARGONNE NATIONAL LABORATORY
APRIL 4–7, 2006

Anatolij Afanasjev
Mississippi State University
afansjev@ERC.MsState.Edu

Sverker Almehed
University of Notre Dame
salmehed@nd.edu

John Arrington
Argonne National Laboratory
johna@anl.gov

A. B. Balantekin
University of Wisconsin-Madison
baha@physics.wisc.edu

Bruce Barrett
University of Arizona
bbarrett@physics.arizona.edu

Ian Bentley
University of Notre Dame
ibentley@nd.edu

Mandar Bhagwat
Argonne National Laboratory
mbhagwat@phy.anl.gov

Roelof Bijker
Instituto de Ciencias Nucleares/UNAM
bijker@nucleares.unam.mx

Arnold Bodmer
Argonne National Laboratory

Mark Caprio
Yale University
mark.caprio@yale.edu

Arthur Champagne
University of North Carolina
aec@tunl.duke.edu

Richard Chasman
Argonne National Laboratory
chasman@phy.anl.gov

Timothy Chupp
University of Michigan
chupp@umich.edu

Fritz Coester
Argonne National Laboratory
coester@anl.gov

Sidney Coon
U.S. Department of Energy
sidney.a.coon@science.doe.gov

Andrew Cumming
McGill University
cumming@physics.mcgill.ca

Pawel Danielewicz
Michigan State University
danielewicz@nscl.msu.edu

David J. Dean
Oak Ridge National Laboratory
deandj@ornl.gov

Willem Dickhoff
Washington University
wimd@wuphys.wustl.edu

Jacek Dobaczewski
Warsaw University
Jacek.Dobaczewski@fuw.edu.pl

J. P. Draayer
Louisiana State University
draayer@sura.org

Thomas Duguet
Michigan State University
duguet@nscl.msu.edu

Jonathan Engel
University of North Carolina
engelj@physics.unc.edu

Henning Esbensen
Argonne National Laboratory
esbensen@phy.anl.gov

Stefan Frauendorf
University of Notre Dame
sfrauend@nd.edu

Chris Fryer
Los Alamos National Laboratory
fryer@lanl.gov

Donald Geesaman
Argonne National Laboratory
geesaman@anl.gov

Yingqiong Gu
University of Notre Dame
ygu@nd.edu

Wick Haxton
University of Washington
haxton@u.washington.edu

W. R. Hix
Oak Ridge National Laboratory
raph@ornl.gov

Roy Holt
Argonne National Laboratory
holt@anl.gov

Hisashi Horiuchi
Kyoto University
horiuchi@ruby.scphys.kyoto-u.ac.jp

Mihai Horoi
Central Michigan University
horoi@phy.cmich.edu

Prashanth Jaikumar
Argonne National Laboratory
jaikumar@theory.phy.anl.gov

Robert Janssens
Argonne National Laboratory
janssens@phy.anl.gov

Aksel Jensen
University of Aarhus
asj@phys.au.dk

Teng-Lek Khoo
Argonne National Laboratory
khoo@anl.gov

Dieter Kurath
Argonne National Laboratory
kurath@anl.gov

Denis Lacroix
Michigan State University
lacroix@nscl.msu.edu

T.-S. Harry Lee
Argonne National Laboratory
lee@phy.anl.gov

C. J. Lister
Argonne National Laboratory
lister@phy.anl.gov

Zheng-Tian Lu
Argonne National Laboratory
lu@anl.gov

Gabriel Martinez-Pinedo
GSI Darmstadt
g.martinez@gsi.de

Bradley Meyer
Clemson University
mbradle@clemson.edu

Witold Nazarewicz
University of Tennessee / ORNL
witek@utk.edu

Thomas Neff
Michigan State University
neff@nscl.msu.edu

Kenneth Nollett
Argonne National Laboratory
nollett@anl.gov

Kaori Otsuki
University of Chicago
otsuki@uchicago.edu

Nidhi Patel
Argonne National Laboratory
npatel@anl.gov

Fang Peng
University of Chicago
fpeng@oddjob.uchicago.edu

Muslema Pervin
Argonne National Laboratory
muslema@phy.anl.gov

Steven Pieper
Argonne National Laboratory
spieper@anl.gov

Michael Ramsey-Musolf
California Institute of Technology
mjrm@caltech.edu

Craig Roberts
Argonne National Laboratory
cdroberts@anl.gov

Vincent Rotival
Michigan State University
vincent.rotival@polytechnique.org

Alfred Sattelberger
Argonne National Laboratory
asattelberger@anl.gov

Guy Savard
Argonne National Laboratory
savard@phy.anl.gov

Ivo Seitenzahl
University of Chicago
irs@uchicago.edu

George Souliotis
Texas A&M University
soulioti@comp.tamu.edu

Jie Sun
University of Notre Dame
jsun@nd.edu

Pieter van Isacker
GANIL
isacker@ganil.fr

Bira van Kolck
University of Arizona
vankolck@physics.arizona.edu

James Vary
Iowa State University
jvary@iastate.edu

Alexander Volya
Florida State University
volya@nucmar.physics.fsu.edu

Robert Wiringa
Argonne National Laboratory
wiringa@anl.gov

Stewart Wright
Argonne National Laboratory
svwright@anl.gov